ROADSIDE GEOLOGY

of New York

Bradford B. Van Diver

MOUNTAIN PRESS PUBLISHING COMPANY
MISSOULA

Sixth Printing, April 1994

Library of Congress Cataloging-in-Publication Data

Van Diver, Bradford B.
Roadside geology of New York.

Bibliography: p.
Includes index.
1. Geology—New York (State)—Guide-books. I. Title.
QE145.V35   1985   557.47   85-13871
ISBN 0-87842-180-7

Printed in the U.S.A.

MOUNTAIN PRESS PUBLISHING COMPANY
P.O. Box 2399 • 1301 S. Thirds St. W.
Missoula, Montana 59806
(406) 728-1900

Dedication

To geology students everywhere

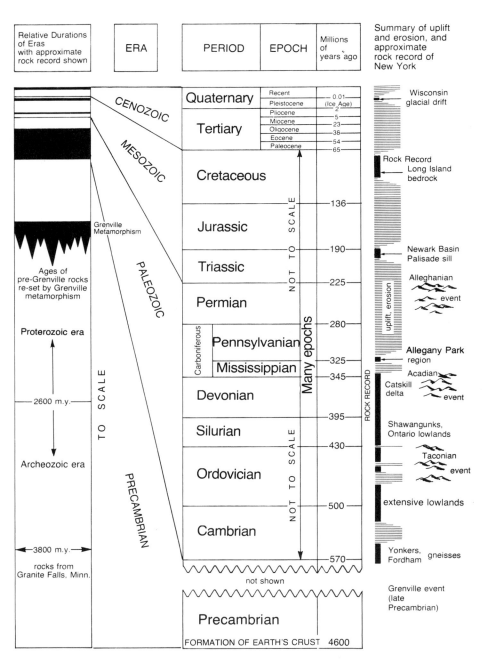

Geologic Time Scale

The column on the left is scaled to the full length of earth history and the
relative durations of the geologic eras, with black bars indicating the preserved
rock record in New York. The right column is not to scale, and shows time
divisions and tectonic history of post-Precambrian time only.

PREFACE

New York is full of exciting geology, and much of it is visible from the highway. That's what this book is all about—what you can see from your car, and where it fits into the geological picture, what it means.

Because there is so much to see, and the state is such a pretty one, we have had a lot of fun doing this. The other person in the "we" is my wife, Bev, who has worked by my side through the whole project. We logged the 4580-plus miles of roads (with about 10,000 actually traveled) together. I did the driving and observing while she did the recording. She made many astute observations, too, which were all the better because she has no training in geology. Through her eyes, I saw more of the things a layman would see and ask questions about. When we came to the writing stage, Bev's years of teaching English were invaluable; she was quick, for example, to spot passages that were unclear and thus might pose a problem for my readers. Bev also typed all of the text on our little computer. Her name doesn't appear on the cover, but there is a lot of her in here, and it is to her that I owe the greatest debt of gratitude.

This book is designed primarily for the layman and geology student. There is a lot of "meat" here, though, that I hope my professional colleagues will find useful. The language is quite plain, as free of geological jargon as I could make it; yet there is a glossary in the back to help with terms that are used. I hope that you find it easy-going, interesting, and informative.

The first thing you have to do to write a book like this is to read as much as you can of what's been written about New York geology, so that when you see a particular roadcut, outcrop, or landscape, you'll have some idea of what's going on. It's a big state; and because of this, it's impractical to expect to go back to certain areas where you didn't get everything you needed the first time. You have to be well prepared to do it right the first time.

As for sources, I hasten to acknowledge the invaluable guidebooks of the New York State Geological Association. NYSGA has been meeting annually since 1925, except for the World War II years, 1942-1945. Each year the meeting is held in a different part of the state, and a guidebook is prepared by the hosting institution for local field trips. I have used many of these guides, and I am indebted to a large number of their contributors. My apologies to them for not mentioning their names individually. Although these guides are mostly written for the professional, you may wish to obtain some of them to get more detailed information about particular areas. You will find them in many libraries across the state.

The second greatest source of information has been the many publications of the New York State Museum and Science Service (The Geological Survey). These include Bulletins, Map and Chart Series, Memoirs, Special Publications, and, especially designed for the layman, The Educational Leaflets and Handbooks. A list of all the above published through 1975, may be found in the Empire State Geogram, vol. 11, No. 2, 1976.

Still more geological background was gleaned from a wide range of journals, particularly those of the United States Geological Survey and the Geological Society of America. Still other sources are listed under *Suggested Readings* at the back of the book.

Many of the photographs and line drawings included herein (some modified) also appeared in my earlier book, *Geology Field Guide to Upstate New York* (1980), published by Kendall/Hunt Publishing Company and used with their permission. The reader will find that book an excellent companion for this one.

A number of highly artistic drawings herein are the creations of one of my geology students at Potsdam, Ms. Theresa

(Terry) Jancek, and I am especially fortunate and grateful to have had her outstanding talent available. One drawing of an Ordovician life scene in the Trenton (limestone) sea was done by another student, Hannan E. LaGarry. Otherwise, the art work is all my own, some of it adapted from other publications, as noted. The photographs are mine, too, with the exceptions of the ERTS-1 satellite mosaic of the Adirondacks, the photo-mosaic of Lake George (by Richard K. Dean of Glens Falls, New York) and the underground photo of Howe Cavern (by Rodney Schaeffer). Richard L. Bitely of Potsdam did the processing.

My thanks also to my geology colleagues at Potsdam: Drs. Jim Carl, Bill Kirchgasser, Frank Revetta, and Neal O'Brien for helpful discussions. Jim edited the Plate Tectonics chapter and made many useful suggestions. Material for this important account comes mainly from "A Brief Tectonic History of New York" by D. W. Fisher, Y. W. Isachsen and L. V. Rickard, that appears with the 1970 edition of the state geologic map. Thanks, too, to Dr. Fred Wolff of Hofstra University who kindly led me in the field to some of the important geologic sites of Long Island.

Each route description, or chapter, is accompanied by a geologic map that is a simplified version of part of the state geologic map. You will find considerably more detail on the map than in the text, but that's as it should be. A picture is worth a thousand words. With only a few exceptions, the maps show bedrock, to the exclusion of unconsolidated, surficial deposits, which in this state are overwhelmingly glacial in origin and are often more visible than the bedrock. It was either the bedrock or the surficial deposits, for to show both would have made the maps incomprehensibly busy. The bedrock exerts the strongest influence, by far, on the development of the large-scale landscape features. The physiographic provinces of the state, for example, are truly also geologic provinces distinguished by unique bedrock types, ages, and geometry. For the larger setting of the roadside geology, you have the two state maps at the beginning of the book, one showing bedrock, the other showing glacial features, and both showing all of the routes described herein to help you get your bearings. Repeated reference to these handy maps and to the Geologic Time Scale, should, in time, enable you to sense the geologic personality and incredibly long history of this most interesting state.

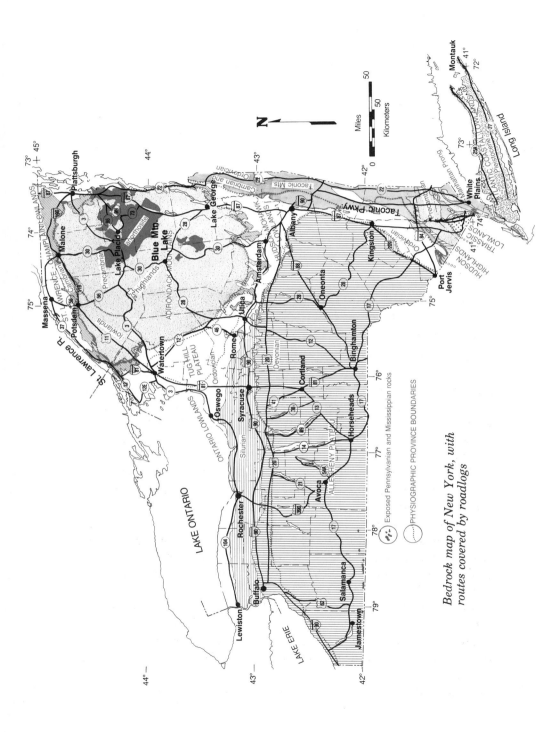

Bedrock map of New York, with routes covered by roadlogs

Exposed Pennsylvanian and Mississippian rocks

PHYSIOGRAPHIC PROVINCE BOUNDARIES

Table of Contents

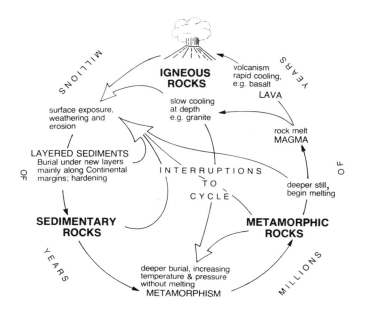

The rock cycle, showing the three major classes of rocks, how they form, and how they interrelate

BEFORE WE BEGIN

A WORD ABOUT ROCKS

New York's ancient geologic history is written in its rocks, as surely as if it were printed; and you don't need to be an expert to read it. For example, just being able to identify the type of rock and knowing how rocks form reveal nearly half of the story. Rocks are always forming somewhere on earth, albeit slowly, while others are being eroded elsewhere, or melted, or metamorphosed—recycled. The rock cycle shows the three major classes of rocks—igneous, sedimentary, and metamorphic; how they form and therefore, what they mean; and how they are interrelated.

The preserved rocks are the rock record. They are like the pages of a book, but there are always pages missing, and the existing pages are often crumpled and shuffled by violent geologic events of the past. The rock record of New York is this way. Nevertheless, several generations of geologists have managed to put all the pieces together into a coherent and fascinating history.

The Geologic Time Scale is the product of this kind of piecing together of rock records, not just from New York, but from the entire world. Obviously, we can only claim a very small part of the total history of earth; information to fill in the enormous gaps, the missing pages, called unconformities, must be obtained from somewhere else where that part of the record is intact. The dating of different rocks in the record has been laboriously achieved by using such evidence as the fossil record, stratigraphic position, and the decay of radioactive elements like uranium that are sometimes contained in rock-forming minerals. Once dated, the rock units are placed in their proper chronologic order—like the pages of a book. Usually, the oldest rocks, and the geologic events they depict, are placed at the bottom of the scale, because that is most often where nature has placed them.

The earth is incredibly old, at least 4600 million years old. It is difficult to comprehend such a large span of time; and we can't relate to it with our human clocks, consisting of days, weeks, months, years—even millenia are too short. The geologic clock ticks away a million years at a time, and the Geologic Time Scale fills the need for a different kind of calendar. Its basic units are called periods. Just as our weeks are lumped together in months, the geologic periods are lumped into large units called eras, and just as we divide weeks into days, periods are split into epochs. The divisions are based on significant events recorded in the rocks: such as mountain-building, submersion beneath the sea, or the rise or fall of certain life forms. As a result, none of the divisions represent the same number of years.

As you travel the highways and byways of New York, let the Geologic Time Scale be your road map to history. Throughout the book, you will repeatedly come across names like Precambrian (actually two eras, the older Archeozoic, and the younger Proterozoic), Cambrian (period), or Pleistocene (epoch). Geologists know these names like the backs of their hands, because they must use them constantly as a time frame. If you keep referring to the time scale when you encounter them, you too will soon become familiar with them and the events they depict.

SOME COMMON ROCKS OF NEW YORK

Here is a reference list to help you identify some of the rocks you will see most frequently along the highways of New York.

A. **Sedimentary rocks:** These are bedded rocks formed by consolidation of layer upon layer of loose sediments. They are especially useful clues to geologic history because they record past conditons at the earth's surface, and often contain fossils.

    1. **Shale:** The single most common rock type in New York. Shale forms from clay and mud. Fine-grained, generally dark-colored, and flakey, it erodes easily to shingled slopes. It is the principal bedrock of the Allegheny Plateau in the so-called Catskill delta where it is locally highly fossiliferous. Shale is also common in the Ordovician, Silurian, and Triassic lowlands.

    2. **Siltstone:** Much of the shale in New York contains siltstone interbeds composed of slightly coarser material that gives it a gritty feel. It may often also be recognized by its thicker bedding which lacks the flakiness of shale, and by its lighter color.

    3. **Graywacke:** This is a common variety of grayish sandstone consisting of mixtures of sand, silt, and clay, and sometimes even gravel. It is often interbedded with shale.

    4. **Limestone and dolostone:** These are shallow sea carbonates that are generally light to dark gray, and thin- to thick-bedded. They commonly form cliffs, and some units are a virtual cornucopia of fossils. They often contain caves. Carbonate rocks form the caprock of Niagara Falls, and the Niagara, Onondaga, and Helderberg scarps. They occur in abundance in the lowlands adjacent to the Adirondacks. They are often interbedded with shales.

    5. **Sandstone:** Ordinary sand hardens into sandstone, a solid rock. It is commonly reddish or yellowish-brown from iron oxide staining, well-bedded and sometimes cross-bedded, and sandy-textured with grains rounded as a result of stream transport. It forms cliffs and ridges. The most famous units are the Cambrian-Ordovician Potsdam sandstone that rims the Adirondacks, and the Devonian "bluestones" of the Letchworth Park region. It generally represents shore or near-shore deposits. It is often interbedded with shale or siltstone which are finer than sandstone.

    6. **Conglomerate:** This is a consolidated gravel-sand deposit with rounded, stream-transported pebbles, cobbles, or boulders enclosed by finer matrix, that is commonly massively and poorly bedded. It forms "rock cities" on hillslopes of the Shawangunks (Silurian) and southwestern New York (Devonian, Mississippian, Pennsylvanian), redbeds in the Triassic Lowlands, and cliffs in the Catskill Mountains (Devonian). Small pockets underlie Potsdam sandstone (Cambrian?) in many places. It represents mountain slope or seashore deposit.

B. **Metamorphic Rocks:** These are "changed" rocks, created primarily when sedimentary, igneous, or older metamorphic rocks are subject to extreme temperatures and pressures in the core of a building mountain range, like the ancestral Adirondacks or the Taconics. Metamorphism results in new mineral makeup and new textures induced without melt-

3

ing. Complex deformation of the original, generally even, layering often occurs.

1. **Gneiss:** This is a coarse-grained, streaky, sometimes banded rock with dark and light layers, that makes up most of the Precambrian Hudson Highlands and Adirondacks and much of the Manhattan Prong. Many of the more homogeneous, non-streaky varieties in the Adirondacks are metamorphosed igneous intrusive rocks, ranging from dark green metasyenite to whitish metanorthosite. It generally represents medium to high temperature metamorphism.

2. **Marble:** This is a generally whitish or grayish, streaky, and coarse-grained rock, often with swirling, very fluid patterns of deformation. It may contain blocks of other rocks that have been swept along with the flow. It forms valleys and low, solution-pocked outcrops, a major rock-type of the Precambrian Lowlands Adirondacks where it is called Gouverneur marble, Manhattan Prong (Inwood marble, Ordovician?), and Taconics (Ordovician).

3. **Schist:** This is a generally dark, sparkly rock that splits along layers of mica, and is commonly highly deformed. Formed by medium temperature metamorphism of shales, it is a common rock of the southern Taconics and Manhattan Prong, less common in the Lowlands Adirondacks.

4. **Slate:** This is a fine-grained, low-temperature equivalent of shale that splits easily into thin, flat, cleavage plates. It is the principal rock-type of the western side of the Taconic Mountains, where it forms dark, crumpled, slabby outcrops.

5. **Phyllite:** This is a slightly higher-temperature equivalent of slate of the eastern Taconics, and distinguished from it by more irregular cleavage surfaces that display a pearly sheen from very fine-grained metamorphically-crystallized mica flakes.

6. **Quartzite:** This is metamorphosed sandstone that is generally harder, glassier, and feels less sandy. It is most easily distinguished from ordinary sandstone by close association with other metamorphic, rather than sedimentary, rock-types.

C. **Igneous Rocks:** These are "melt" rocks formed by crystallization of lava or its subsurface parent material, called magma, that may originate either from the earth's mantle or as an end-product of extreme metamorphism in the deeper recesses of the crust.

1. **Basalt or diabase:** This is a dark gray to black, dense rock that commonly weathers to a rusty color, and is usually highly fractured. The best example is the Palisades Sill which "intrudes" between the layers of Triassic sedimentary rocks and is exposed in the New Jersey Palisades across the Hudson River from Manhattan and the Bronx. It is also common as dikes, tabular intrusions, that cut across the layering of the host rock in the Adirondacks. It represents near-surface intrusions during volcanic episodes.

2. **Granite:** This is a generally light-colored, coarse-grained rock formed by extremely slow crystallization of magma deep within the core of a mountain range. The Devonian Peekskill granite is the best example. There are, however, many rocks in the Adirondacks and Manhattan Prong that resemble igneous granite, but are metamorphic, and are referred to as granitic gneiss.

4

D. **Other rocks:** Some important rocks have been deliberately omitted from the detailed discussion above, for various reasons. The extensive Silurian rock salt and gyprock beds of western New York, for example, are seldom seen on the surface because they weather so easily. One of the important Adirondack rock-types is metanorthosite, which will be treated separately in route descriptions. The Adirondacks, in fact, contain a wide variety of metamorphic rocks that have been lumped as "gneiss," in the interest of simplification. The Cortlandt igneous complex near Peekskill contains a host of unusual, dark-colored, coarse-grained rocks, called "ultramafics"; and another ultramafic called serpentine is a significant member of the rock suite along the "suture line" of southeastern New York.

E. **Non-rocks or unconsolidated sediments:** The unconsolidated glacial deposits of New York demand attention because they are nearly ubiquitous and very visible. The various kinds of deposits are often lumped together as glacial drift, and subdivided into unstratified and stratified drift. The former generally represent sediments simply dumped in place by melting of the enclosing ice and the latter are meltwater stream or lake deposits.

1. **Till:** This is unsorted, unstratified drift that varies widely in composition and character, but may contain boulders of enormous size, as well as glacial flour consisting of ground-up rock material. It is found in stony fields, drumlins, moraines (glacier margin deposits) such as the Ronkonkoma and Harbor Hill moraines of Long Island and the Valley Heads moraine of the Finger Lakes region.

2. **Outwash:** This is sandier, more stratified, better sorted material that has been transported by meltwater streams. It is found in outwash plains like those of the southern half of Long Island, and those south of the Finger Lakes downstream from the Valley Heads moraine. The term may also be loosely applied to esker and kame sediments.

3. **Varves:** These are grayish, banded, clay-silt deposits formed by slow sedimentation in glacial meltwater lakes. They generally occur in thick sequences of horizontal, dark-light, paired laminations, with each pair representing one year's sedimentation (light-summer, dark-winter).

LOOKING AT THE LANDSCAPE

Geology and topography are inextricably tied to each other, perhaps displayed no better than in New York State. It is partly because of this intimate relationship that New York is such a fine candidate for *Roadside Geology*. Knowing something about landscapes, bedrock geology, and geologic processes enables the traveler to see beyond the mere surface features, to know something of what lies beneath, to read geologic history, and to experience the excitement of discovery.

The geology of New York is varied, and the state is neatly divided into provinces, each of which displays its own unique brand of geology, physiography, and scenic pleasures.

The Adirondack Mountains, for example, constitute the state's highest and most rugged landscape. Their rocks are hard, tough gneisses that have been intricately and complexly deformed. At 1.1 billion years, they are so ancient that one would expect them to be completely worn down by now. A clue as to why they are not lies in the oval outline of the province, and the slight uptilting of much younger, layercake sedimentary rock beds around their edges. There is now little question what these features mean: the mountains have been rising like a giant dome, or overturned bowl, in geologic times so recent that they have not yet had time to wear down. In truth, they are still rising! Recent re-leveling surveys by United States Coast and Geodetic Survey crews revealed that they have been rising in modern times at the rate of about 3 millimeters per year. That seemingly miniscule rate is measureably much faster than the rate of erosion, and rapid enough to raise the roof nearly 3 kilometers in the geologically short period of a million years!

The picture, then, is one of an ancient metamorphic rock terrane being leveled, or nearly leveled, by long-term erosion, and then being covered by the sea. Layer upon layer of sediments were deposited and later consolidated to the same sedimentary rocks we see around the edges of the Adirondacks

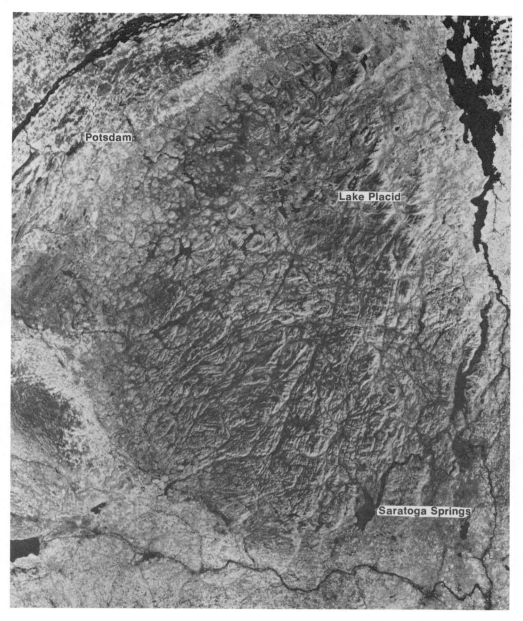

Infrared photomosaic of the Adirondacks region assembled by Yngvar W. Isachsen of the New York State Geological Survey from ERTS-1 satellite images; reproduction by U.S. Geological Survey.
Courtesy Kendall/ Hunt Publishing Co.

today. In an oversimplification, much later, hundreds of millions of years later, the doming began, as though a giant were punching his fist up under the sedimentary strata. The stage was set for erosional stripping of the stratified caprock and deep dissection of the underlying complex metamorphic rocks that produced today's landscape.

Brittle fracture map of the Adirondack region, including up-down block faults and joints. Fracturing diminishes to northwest and has a dominant northeast trend. Several graben contain Ordovician strata, indicating probable Taconian age for fracturing.
Adapted from Isachsen and McKendree (1977).

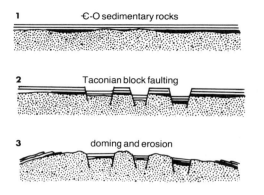

Hypothetical sequence of Cambrian-Ordovician sedimentation (1), Taconian block-faulting (and erosion) (2), and doming (and erosion) (3) of the Adirondacks, leading to preservation of Cambrian-Ordovician strata in down-dropped graben

The lowlands country surrounding the mountains, include the St. Lawrence Lowlands, the Champlain Lowlands, the Hudson-Mohawk Lowlands, and the Erie-Ontario Lowlands provinces. To be sure, there are differences between these regions, but they all consist of low, fairly even topography, built upon nearly flat-lying Cambrian and Ordovician sedimentary strata (and Silurian in the western New York part of the Ontario Lowlands), much of it erosionally weak. Again, geology is, so to speak, the mother of topography.

The Allegheny Plateau province of western New York is a small part of the much broader Appalachian Uplands province. The topography in this region, despite the plateau label, is really of a slightly different character, and the difference is important. The layercake sequence of shaley Devonian sedimentary rock strata dip, or slope, slightly to the south so as to form a tilted mesa, or cuesta. Long-term erosion has carved a step-like landscape on this sequence, with north-facing scarps carved from the upturned ends of the more resistant beds of limestone or sandstone, and broad "treads," or dip slopes extending southward over these same caprock layers.

Don't expect anything too dramatic, however, for this tilted mesa character is subtle and extensively camouflaged by stream and glacial dissection of the plateau. The province includes the Catskill and Helderberg mountains at the eastern end, and extends all the way to the western tip of the state, including the famous Finger Lakes. The boundary with the lowlands nearly everywhere coincides with the boundary between strata of Silurian and Devonian ages.

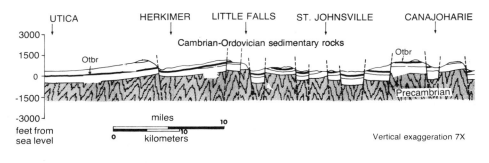

3000 —
1500 —
0 —
-1500 —
-3000 —
feet from
sea level

UTICA   HERKIMER   LITTLE FALLS   ST. JOHNSVILLE   CANAJOHARIE

Cambrian-Ordovician sedimentary rocks

Otbr

Otbr

Precambrian

miles   10
0   kilometers

Vertical exaggeration 7X

W-E structure section

The Tug Hill Plateau province is not much more than an outlying remnant of the Allegheny Plateau, separated from it by a narrow strip of Ontario and Mohawk lowlands, and from the Adirondacks, by the Black River Valley, an arm of the Ontario Lowlands. It is made up of Ordovician, rather than Devonian rocks, and capped by resistant Oswego sandstone.

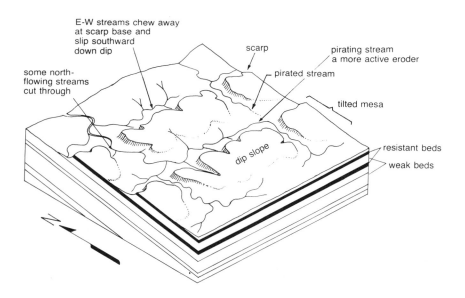

E-W streams chew away at scarp base and slip southward down dip

some north-flowing streams cut through

scarp

pirating stream a more active eroder

pirated stream

tilted mesa

dip slope

resistant beds

weak beds

Tilted mesa, or cuesta, the dominant geologic/topographic character of the Allegheny Plateau region. In fact, this produces the east-west belts of Paleozoic rocks throughout western New York. Courtesy Kendall / Hunt Publishing Co.

10

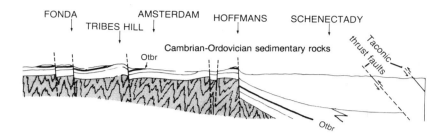

FONDA   AMSTERDAM   HOFFMANS   SCHENECTADY
    TRIBES HILL

Otbr

Cambrian-Ordovician sedimentary rocks

Taconic thrust faults

Otbr

Mohawk Valley   Broughton et al 1966

The rocks dip gently away from the Adirondack dome, so that the principal scarp is on the Black River side. The slope there, and on the north, descends in steps capped by Trenton limestone beds. If anything, the tilted mesa landscape of the Tug Hill is even more conspicuous than that of western New York.

The close relationship between geology and topography is equally well demonstrated in the narrow Taconic Mountains subprovince of the New England Uplands, along the eastern margin of the state, adjacent to Connecticut, Massachusetts, and Vermont. The mountains are largely composed of metamorphic slates and schists that have either been pushed westward into the state along low-angle thrust faults, or have slid, in gigantic landslides, down the slope of an ancestral, Himalayan-scaled mountain range of which the modern Berkshires of New England are only the eroded stumps. In either case, the entire mountain mass is referred to as the Taconic Klippe, literally 'a relocated piece of crustal rock, in this case a very large piece. Such transportation has crumpled the thinly-laminated rocks and imparted to them a strong north-south structural grain that controls the topography. Valleys are mostly underlain by less resistant marbles. Though linear, the landscape displays a distinctive hummocky surface texture that is a reflection of—and a clue to—the complex tight folding and faulting of the bedrock. The Taconics extend north from the Hudson Highlands in Putnam County to Ticonderoga at the northern tip of Lake George, and include the famous Slate Belt around Granville. Throughout the range, the metamorphic grade increases slightly from west to east, meaning that

11

the temperatures and pressures of metamorphism increased in that direction. At the southern end, the gradation is much greater, going from low grade slate in the west to high grade sillimanite gneisses only a few miles to the east. The rocks of the range are of Cambrian and Ordovician age.

The Hudson Highlands are another subprovince of the New England Uplands that form the most prominent topographic feature of southeastern New York. Largely made up of hard gneisses generated 1100-1300 million years ago during the Grenville Orogeny, as in the Adirondacks, the Highlands are part of a northeast-trending mountain mass that extends from Pennsylvania to Connecticut. The many long, straight valleys result from deep erosion of fault or fracture zones. The Hudson River has cut a narrow, 15-mile long gorge through the range between Peekskill and Newburgh that served as a channelway for ice erosion during Pleistocene glaciation. The gorge is a true fjord, like those of the Norwegian coast, a glacially-gouged valley now invaded by the sea, and through which daily tides reach 160 miles inland to Troy!

The remaining subprovince of the New England Uplands in New York State is called the Manhattan Prong. It lies in a narrow strip of New York east of the Hudson Highlands and the Hudson River, and underlies Manhattan and Staten islands. Again, geology controls topography, with low, northeast-trending ridges carved from resistant gneisses and schist, and shallow valleys from weaker marble. An unusual aspect of this region is the exposure of a suture line in Staten and Manhattan islands, the Bronx, and Westchester County, where volcanic islands that once existed offshore were shoved against the continent and welded to it.

Sandwiched between the Hudson Highlands and Manhattan Prong, and lying entirely within Rockland County is the Triassic Lowland province. This region is made up of redbeds of shale, sandstone, and conglomerate, and the Palisades sill, an igneous rock which, as magma, worked its way in between the sedimentary beds. Unlike most of the other sedimentary rocks of the state, the redbeds are largely continental, or land, deposits rather than marine. The lowland on which they rest results from deep erosion of soft rocks accomplished primarily during the first 30 million years of the Triassic Period. The redbeds were formed in late Triassic time as alluvial fan deposits of gravel, sand, and mud carried by fast-moving streams from the bordering highlands. The Palisades sill forms a prominent, columnar facade across the Hudson River from Manhattan, Yonkers, and Westchester County.

Long Island is part of the broader Atlantic Coastal Lowlands province that stretches north and south along the east coast. It is not much more than a ridge, or blanket, of direct glacial and glacial outwash sediments that almost completely conceal the underlying Cretaceous sedimentary bedrock. Long Island topography, therefore, is glacial topography, with little or no influence from the underlying, erosionally-reduced Cretaceous strata. The Long Island shoreline displays some of the most dynamic geology of the state, where catastrophic changes can occur virtually "in the blink of an eye."

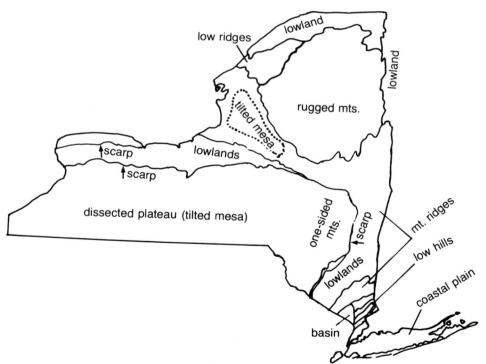

Physiographic elements of New York

I

PLATE TECTONICS

NEW YORK AND THE DRIFTING CONTINENTS

Thus far, we have considered only the relatively small-scale geologic features within the state of New York: the rocks, the landforms, and the geologic/physiographic provinces that are unique combinations of rocks and landforms. Now let us look at some really large-scale geologic phenomena to see how New York got the way it is. To do this, we must refer to a new geologic frame of reference called plate tectonics.

Plate tectonics is a revolutionary concept that had its real beginning in the 1960s and has been snowballing ever since. It is rooted, however, in the old idea of continental drift that was proposed by the German meterologist Alfred Wegener early in the 20th century. According to the modern concept, the outer thin shell of the earth, called the lithosphere, is broken into a small number of plates of enormous size (perhaps 12) resembling slabs of Arctic sea ice. Like sea ice, the plates are constantly moving about, albeit much more slowly, causing some to collide, some to rift apart, and some to grind past each other. Also like sea ice, the crustal slabs "float" on a denser layer beneath. This underlayer, called the asthenosphere, is solid rock that behaves plastically because of the high pressure and temperature associated with deep burial. The driving force for the movement is believed to be radioactive heat released within the earth's interior, which produces giant convection cells like those in a boiling pot of water. Where cells well up under the lithosphere and diverge, the plates pull apart. Where cells converge and plunge back to the depths, plates collide.

Most plates include segments of both continental and oceanic crust. For example, the North American plate encompasses North America, Greenland, half of Iceland, and half of the north Atlantic oceanic crust. The South American plate includes all of that continent and half of the south Atlantic. The African plate is much larger and includes large portions of the ocean basins on all sides of the continent.

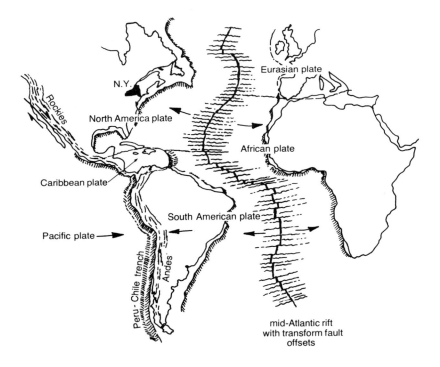

Major plate tectonic elements of the western hemisphere as they appear today. The Atlantic Ocean basin is still actively spreading as it has been for about 200 million years, shoving North and South America against the Pacific plate, producing an oceanic trench, mountain-building, and volcanism.

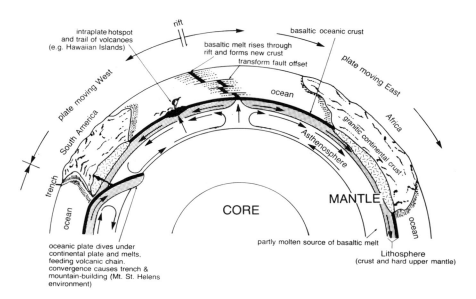

Schematic cross-section showing the major subsurface plate tectonic features of the south Atlantic Ocean as they exist today; the North Atlantic is essentially the same. Courtesy W. H. Freeman Publishing Co.

A prime example of a pull-apart plate boundary is the great submerged chain of volcanic mountains that runs the length of the mid-Atlantic seafloor. This mountain chain, called the mid-Atlantic ridge, has a deep rift valley along its crest. The mountains are volcanic and formed by upwelling of lava through cracks generated when plates on either side of the rift pull apart. The lava cools and hardens to form new oceanic crust. Such new crust constitutes Iceland, one of the few places where rift lava erupts above sea level. Elsewhere, the lava crust is almost entirely submerged.

By contrast, Mt. St. Helens and the other Cascade volcanoes of the west coast lie along a converging plate boundary where the more dense basaltic Pacific plate dives under the edge of lighter granitic rock of the North American continent. Parts of the descending crust are melted at depth to provide new magma for the volcanic eruptions.

The San Andreas fault is an outstanding example of the third type of boundary, where plates slide past each other. A series of jerks and pauses accompanies the movement of the plate containing western California which is slowly drifting northward past the rest of the state and the rest of the North American plate. The fears of some Californians that a piece of their earthquake-wracked state may one day push off to sea have some basis in fact! That "one day," however, will almost certainly happen gradually for millions of years.

Plate tectonics has been going on for a very long time, perhaps as much as half of earth history. As a result, the positions, sizes, and shapes of the continents and ocean basins and all of their associated features have been and still are in a state of perpetual change. Ocean basins have repeatedly opened and closed; as they closed, mountain ranges accompanied by volcanism were pushed up along the continental margins. Where continents collided and the oceans closed completely, mountains similar to the modern Urals or Himalayas rose along the suture.

Against this background, we now consider the plate tectonics history of New York. Most of the state is underlain by a "basement" of Precambrian rocks that is physically continuous with rocks in the Canadian Shield, but mostly buried under younger stratified rocks. The shield is a vast terrane of crystalline

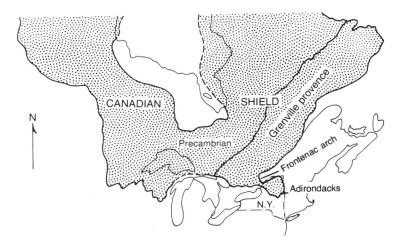

Map of the Canadian Shield, the vast terrane of ancient Precambrian bedrock that forms the nucleus of North America; the Adirondack connection with the Grenville Province of the shield is the Frontenac arch, exposed in the Thousand Islands. Courtesy Kendall / Hunt Publishing Co.

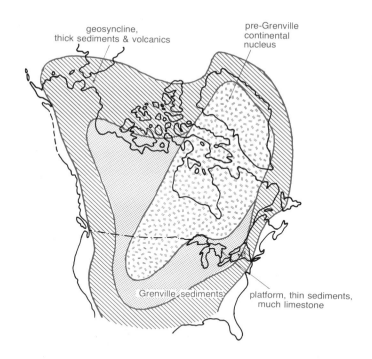

geosyncline,
thick sediments & volcanics

pre-Grenville
continental
nucleus

Grenville sediments

platform, thin sediments,
much limestone

Hypothetical paleogeologic map of North America in late pre-Grenville time. Note that the continental shelf (or platform), with its limestone deposits, extends through northwest New York, where much marble, the metamorphic equivalent of limestone, is found today.

igneous and metamorphic rocks that forms the nucleus of North America and is exposed in about two-thirds of Canada.

Geologists have dated the igneous and metamorphic events in numerous parts of the shield by using the radioactive elements in the rocks. The youngest part of the Canadian Shield, called the Grenville Province, lies along the eastern edge of Canada. It occupies a northeast-trending strip several hundred miles wide extending for nearly 1300 miles from Lakes Huron and Ontario to the northeast coast of Labrador. New York's exposed crystalline basement rocks are an extension of this Grenville Province; they include the Adirondacks, Thousand Islands, Hudson Highlands, and the Fordham gneiss, all of which range in age from 1300-1100 million years. Also included are outcrops in the Berkshire Hills just over the New York border in Massachusetts, and portions of their northern extension, the Green Mountains of Vermont.

Many of the Grenville rocks had their beginning as sand, lime mud, and clay deposited in seas adjacent to the continent. This pre-Grenville period spanned perhaps hundreds of millions of years, when continental and oceanic crusts were locked together. The coastline then must have been a painfully desolate place: flat, devoid of land plants or animals, and a nearly lifeless sea except for mounds of blue-green algae, or stromatolites, poking their cabbage-like heads above the tidal flats. There were undoubtedly some coastal barrier islands like those of the modern south shore of Long Island; intervening lagoons

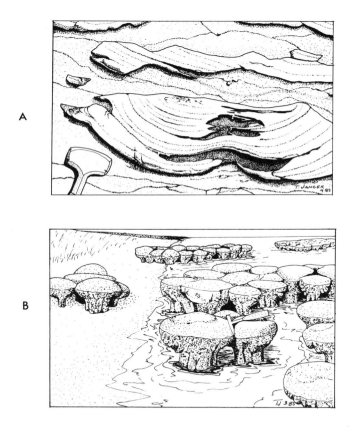

A. Cross-section drawing of an actual stromatolite "cabbage head" found in the 1.1 billion-year-old Gouverneur marble at Balmat, New York, this one overturned by structural deformation. B. Modern stromatolite "cabbage heads," formed by blue-green algae in tidal flats, are periodically exposed to the air at low tide. Drawings by T. Jancek

accumulated evaporite deposits of salt and gypsum. All of the sediments were gradually converted to rock, and, in time, the sedimentary rock strata built up to tens of thousands of feet thick.

The ocean basin began to close. Volcanic eruptions added their materials to the sedimentary pile. By about 1300 million years ago the oceanic crust bordering the early continent was completely consumed, and the continents slowly collided, creating a lofty mountain range stretching from Labrador to Mexico that we'll call the ancestral Adirondacks. Rocks formed on the margins of both of the continents that bordered the closing ocean were crumpled, metamorphosed, and partly melted at great depth.

The continents must have enjoyed a compatible relationship because they did not separate until about 650 million years ago. By then, as much as a 15 mile thickness of rock had been eroded from them, and the mountains were reduced to a surface of little relief. The original mountains were never that high above sea level. To explain, most great mountain ranges float like icebergs, with deep roots that project into the denser rocks below. Like icebergs, they float higher when weight is removed from their tops by erosion. It is the "root rock" that is exposed at the surface today.

The new opening initiated not the present Atlantic ocean but an earlier version called the proto-Atlantic. A new plate tectonic cycle was begun that would involve, as before, long term

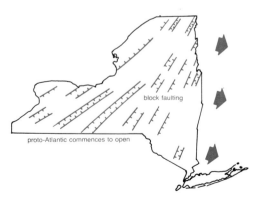

650 million years ago

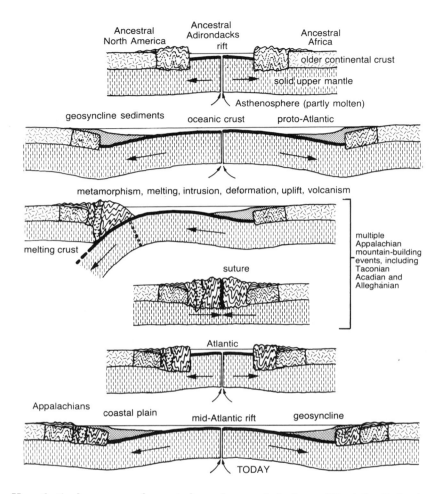

Hypothetical sequence of repeated openings and closings of the Atlantic basin leading to accretion and growth of the North American and African continents

Adapted from Dietz, R. S., 1972, Scientific American, 226, No. 3, 30-38.

erosion and sedimentation on the continental margins as the ocean basin widened, followed by a closing, with accompanying volcanism and mountain-building.

The cycle was interrupted about 575 million years ago by another, incomplete plate covergence which geologists refer to as the Avalonian mountain-building event. The event, however, is poorly understood; and in New York it apparently caused deformation and metamorphism only in the Precambrian rocks of the Hudson Highlands and the Manhattan Prong.

22

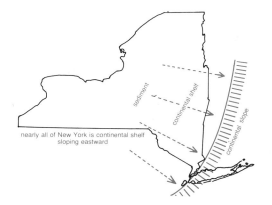

650-445 million years ago

The proto-Atlantic Ocean continued to widen through the late Cambrian and early Ordovician periods as the New York landscape was eroded to a low level. The Atlantic and North American crusts were tightly locked as parts of a single tectonic plate. The state was gradually submerged beneath the shallow, westward-transgressing Potsdam Sea, in which early Paleozoic life forms eventually flourished. In this period, nearly all of the state served as part of the continental shelf upon which great thicknesses of sediments accumulated. The boundary between continental shelf and slope probably passed through the New York City area; thus, the future site of Long Island received the even thicker deposits of the continental slope.

Finally, driven by a reversal of convection currents in the earth's mantle, the proto-Atlantic began to close about 445 million years ago in middle Ordovician time, and the Taconian mountain-building event commenced. At first, the compression appears not to have broken the crust right along the continent-ocean crustal bond. Instead, a rupture occurred offshore, and the crustal shortening was accommodated by shoving of the western segment beneath the eastern segment. The underthrust slab eventually partly melted; and a volcanic island arc formed like those of the western Pacific today (Japan, Phillipines, etc.) Volcanic materials and black muds were added to the sedimentary pile of the continental margin, even

23

as the ancestral Taconic Mountains began to rise. By late Ordovician time, about 435 million years ago, these islands were shoved against and welded to the continent; and the ocean basin was completely consumed. The westward-directed thrust faults, massive landslides, intense folding and fracturing, and metamorphism we see in the modern Taconic Mountains and the lesser deformation west of them, at least as far as the Allegheny Plateau, are the products of the Taconian mountain-building event. So, too, are the Cortlandt and Croton Falls igneous intrusive complexes near Peekskill.

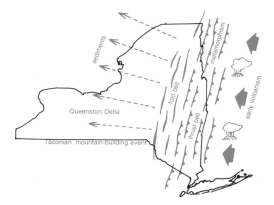

445-435 million years ago

It is difficult to say just how high the ancestral Taconics were, but it is certain that the metamorphism evident in the modern mountains occurred underneath several miles of rock. In other words, an enormous amount of erosion has taken place since then. The sediments thus generated spread westward, displacing a shallow sea and forming an immense apron, named the Queenston Delta from the late Ordovician Queenston formation exposed in the Ontario Lowlands. The Taconian event, therefore, produced not only high mountains, but also completely reversed the sedimentation patterns in New York, from slow, east-directed deposition on a gently sloping continental shelf to rapid, west-directed sedimentation from the mountains toward the continental interior.

Near the end of the Taconian event, the delta plain was, itself, uplifted and subjected to significant erosion, producing

435-425 million years ago

the Taconic unconformity, a gap in the geologic record. Silurian and early Devonian sediments subsequently covered the erosional surface, with the white, quartz-pebble, Shawangunk conglomerate at the bottom of the pile in eastern New York.

This cycle of erosion and deposition marked a renewed crustal stretching of uncertain origin which, in its early stages, caused the extensive up-down, or block, faulting presently so visible in the Adirondack, Mohawk, Hudson, and Champlain regions. Some of the Ordovician strata are preserved on the floors of down-thrown blocks, or graben, where they have been protected from erosion.

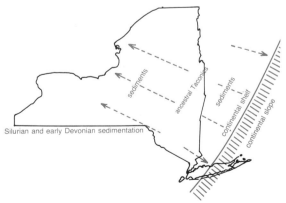

430-375 million years ago

The Acadian mountain-building event, between 375 and 335 million years ago, marked the final closing of the re-opened proto-Atlantic, continental collision, and a much loftier range of mountains in New England, east of the worn-down Taconics in New York. Deformation associated with the uplift was over-printed on all rocks at least as far west as the Adirondacks, including those of Precambrian to Ordovician age that had already crumpled in the Taconian event. Acadian metamorph-ism, however, only reached the eastern edge of the state. Silu-rian and early Devonian strata overlying the Taconic uncon-formity were also tilted westward and mildly folded during this event. In addition, the entire Hudson Highlands block may have been shoved bodily northwestward in response to the compression.

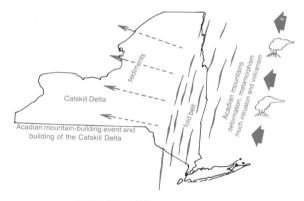

375-335 million years ago

Certainly the most profound Acadian effect in New York was the building of another great delta like the Queenston Delta, as a result of the erosion of the Acadian Mountains, even as they rose. Called the Catskill Delta for exposures in the Catskill Mountains, this enormous blanket of sediments spread west-ward, displacing a shallow inland sea as before and completely burying the stumps of the ancestral Taconics. The consolidated remains are the middle and late Devonian strata of the entire southern tier of New York, the Allegheny Plateau. The Catskill Mural Front is the erosionally-truncated eastern edge and thickest remaining section of the Catskill Delta. To the east, the delta, and even the Acadian Mountains themselves, have since been removed by erosion.

The Peekskill granite and smaller igneous bodies in Westchester County were intruded in post-Acadian time between 335 and 320 million years ago, probably during another period of crustal divergence.

The Alleghanian mountain-building event about 250 million years ago marked a final crustal convergence, but its effect on New York is virtually unknown. Some geologists consider the broad, gentle, east-west folds of the south-central section of the Allegheny Plateau to be Alleghanian. The results of the Taconian, Acadian, and Alleghanian events are all part of the Appalachian foldbelt that now stretches from Newfoundland to the Mississippi basin, and beyond.

The global picture in Alleghanian time is one of a super continent, called Pangaea, consisting of all the present continents fitted together like a jigsaw puzzle and surrounded by a super ocean. Pangaea began to break up about 200 million

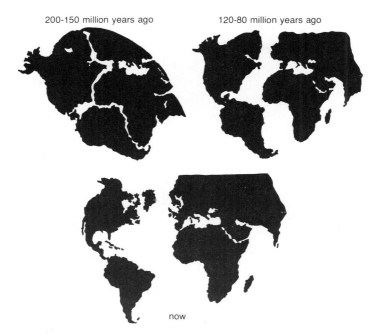

200-150 million years ago   120-80 million years ago

now

Breakup of Pangaea and continental drift in the western hemisphere over the past 200 million years, leading to the present geography

27

years ago to form the modern continents, and a juvenile Atlantic Ocean began to appear in the rift between Africa and Europe on one side and the Americas on the other. The Triassic basins of the east coast, including the Newark basin of New York's Rockland County, dropped along block faults and filled with sediments in the early stages of plate separation at the conclusion of mountain uplift. The Atlantic Ocean basin widens even today, continually splitting apart along the mid-Atlantic rift system at an average rate of about 2.5 cm. per year. Meanwhile, the continental and oceanic crusts on either side of the basin remain tightly locked and inactive; and a new sedimentary pile builds up on the continental margins.

As you can see, New York has been in the "front line" throughout much of the plate tectonic history of North America, and this is precisely why it is such an interesting state geologically. A lot has happened here. There is a consistency to the plate tectonic cycles, despite many differences in detail. Each involved the building of a mountain chain along the edge of the continent, with all of its associated deformation, metamorphism, and igneous activity. Each added new, hard rock crust, so the continent got larger as its margin moved eastward; it was the first (well-documented) cycle, the Grenville event, that built the foundation of the state. Each involved block faulting in response to later plate divergence, and a long period of erosion during which the mountains were reduced to a low level and a new sedimentary pile built up on top of the basement rocks formed earlier. The consistency and the number of cycles over more than one billion years of earth history, leave little doubt that we are even now in the midst of another cycle. The question, it seems, is not whether history will repeat itself again—but WHEN.

The modern New York landscape is largely the product of slow erosion since Alleghanian time, including glacial erosion during the Ice Age. The doming of the Adirondacks and their consequent erosion to the present mountain landscape, probably began only a few million years ago. The uplift is now thought to result from a different kind of plate tectonic activity, a slowly rising plume of hot mantle rocks under the Adirondack crust.

II

THE ICE AGE

NEW YORK GLACIATED

To better comprehend the glacial geology of New York State, we need to review some of the history of the Pleistocene (Ice Age) epoch of geologic time. The Ice Age in North America (and concurrently in Europe) began about two million years ago and ended about 6000 years ago. This long episode was marked by at least four major glacial advances and retreats triggered by fluctuations in the Pleistocene climate. Erosional and depositional evidence for each major advance may be found in other northern states, but in New York only features of the latest or Wisconsin glaciation are well preserved. The clues to earlier advances have been swept away or obscured by this final surge of ice. Earlier glacial deposits are exposed, for example, in Pennsylvania only a few miles south of the southwest corner of New York.

Wisconsin glaciation climaxed about 20,000 years ago when ice sheets covered nearly all of Canada and extended south-

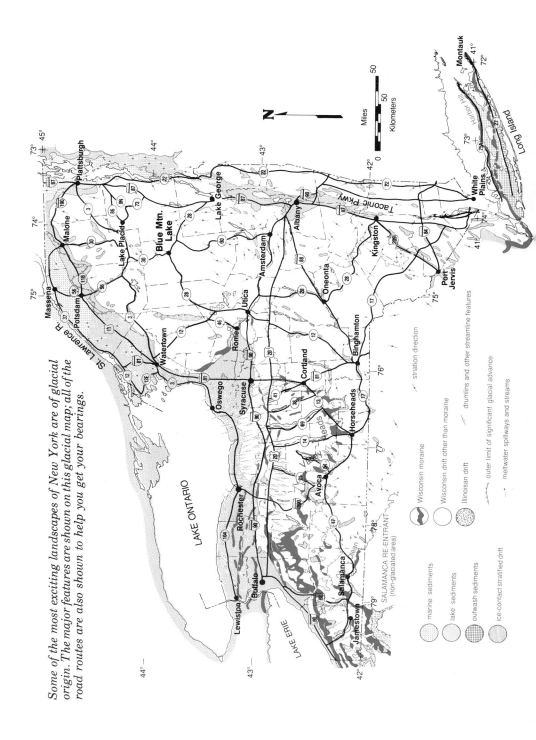

Some of the most exciting landscapes of New York are of glacial origin. The major features are shown on this glacial map; all of the road routes are also shown to help you get your bearings.

30

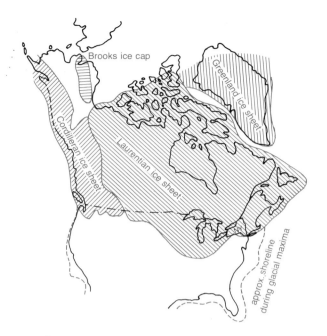

Approximate maximum extent of the major ice sheets in North America during the Ice Age. At climax, so much water was locked in the ice that sea level was as much as 300 feet lower and the coastline far out from its present position, as shown by the dashed line.

ward over the sites of Seattle, Chicago, and New York City. All of New York, except for the Salamanca re-entrant in the Allegany State Park region, lay buried under ice that was more than a mile thick in places. The glacier overrode the Adirondacks, softening the topography and stripping off the soils. Even the highest peaks, Mt. Marcy and Algonquin Peak, were covered.

The Finger Lakes are spectacular testimony to this monumental event. The lakes occupy "troughs," first carved by streams, then gouged out by creeping tongues of ice and finally dammed in the south by a glacial dump pile called the Valley Heads moraine. The patchwork line of the moraine in western New York marks the Wisconsin ice front as it stood about 14,000 to 12,000 years ago. The region between the Finger Lakes and Lake Ontario comprises one of the largest and most striking drumlin fields in the world. Drumlins are hills of glacial debris that have been molded into streamlined forms by overriding ice. They are elongated in the direction of ice movement. Their alignments show that the Lake Ontario basin served as a spreading center for ice that moved radially outward in the region between Buffalo and Watertown. The drumlinoid landscape between Syracuse and Rochester is one of smoothly rolling hills and valleys with a gentle, quiet beauty.

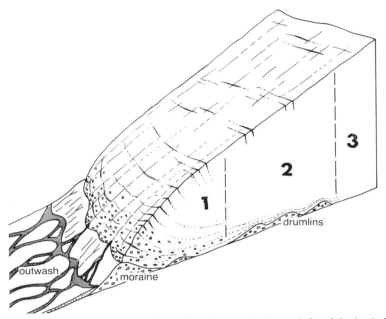

Conditions for the formation of drumlins. In zone 1, the weight of the ice is low and sediments are simply melted out and dumped in place to form moraine. In zone 2, the weight of the ice is just right to cause plastic flow in till and shape it into drumlins. The excessive weight of the ice in zone 3 either causes the till to be bulldozed away or smeared along the ground.
Courtesy Kendall / Hunt Publishing Co.

At climax, Wisconsin ice reached to Long Island. The island is not much more than an enormous sand and gravel deposit. The Harbor Hill moraine forms the northern fluke of the island, and the older Ronkonkoma moraine forms the southern fluke. The rest of the island is largely covered with so-called outwash, sand and gravel carried southward by meltwater streams that washed off the ice and over the moraines.

Morainal topography everywhere in the state tends to be very pitted, owing in part to the presence of kettle holes. These are the sites of former ice blocks left behind by the wasting ice

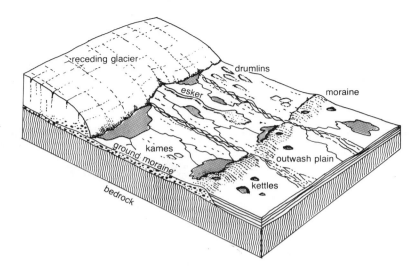

Schematic diagram of a receding glacier and some of the geologic features it leaves behind

Sand and gravel quarry in glacial outwash, showing sorted and cross-bedded character similar to sediments found along modern streams or in deltas
Drawing by T. Jancek

Clusters of glacial kames like these at Mendon Ponds County Park south of Rochester often present a hummocky topography.

that were partially or totally buried in outwash and lake sediments. When they finally melted, they left in their places kettle hole depressions or kettle lakes. The peculiar origin of such lakes is often revealed by their rounded outline and poorly integrated feeder and outlet streams. Morainal areas may also contain kames composed of stratified sand and gravel deposited in deltas along the ice margins by meltwater streams flowing on the ice. They often occur in hummocky kame fields, which differ from moraine in the cleaner, sandy, water-washed and stratified character of their materials. Moraines, by contrast, often contain a wider range of materials and sizes, often in chaotic arrangement, and with a lot of clay; they are commonly referred to as "boulder clay." Another feature frequently found among moraines and kames is the esker, a long, low snake-like ridge of sand and gravel. These are deposits of meltwater streams which flowed under the ice; they were dropped to the ground when the ice melted. Retreat of the continental ice sheet from the Adirondack did not mark the end of all glaciation there. Ice lingered on the high peaks as local glaciers that gouged out valleys of preglacial streams. These valleys now have smoothly-curved floors and walls, and bowl-shaped amphitheaters called cirques at their heads.

As the Pleistocene epoch drew to a close, numerous temporary lakes were formed between the continental ice sheet and high ground. They were the receiving basins for the copious meltwater and for sediments from the ice and barren land. The spillover outlets for these lakes, which controlled water levels,

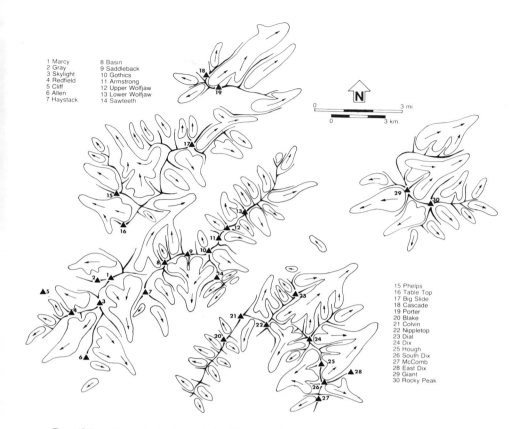

1 Marcy
2 Gray
3 Skylight
4 Redfield
5 Cliff
6 Allen
7 Haystack
8 Basin
9 Saddleback
10 Gothics
11 Armstrong
12 Upper Wolfjaw
13 Lower Wolfjaw
14 Sawteeth

15 Phelps
16 Table Top
17 Big Slide
18 Cascade
19 Porter
20 Blake
21 Colvin
22 Nippletop
23 Dial
24 Dix
25 Hough
26 South Dix
27 McComb
28 East Dix
29 Giant
30 Rocky Peak

Possible valley glaciation of the High Peak region of the Adirondacks in the late stages of Wisconsin deglaciation after the continental ice sheet had backed off the mountains. Map was constructed from 15-minute topographic quadrangle maps by outlining all apparent glacial cirques. All peaks shown are over 4000 feet above sea level. Lower reaches of the glaciers are speculative.

changed frequently as the ice receded. Each time a new spillway was uncovered at a level lower than the existing one, the lake level dropped and its shoreline changed accordingly. Glacial geologists have assigned numerous names to these lakes. Some of the better-known large lakes that are referred to frequently in the following pages are Lake Iroquois (in the Ontario basin), Lake Vermont (in the Champlain basin), and Lake Albany (in the upper Hudson and lower Mohawk lowlands). Shorelines of these and many lesser proglacial lakes are often marked by "hanging" beaches, wavecut cliffs and benches, river deltas, dunes, and other features that have counterparts along modern lakeshores.

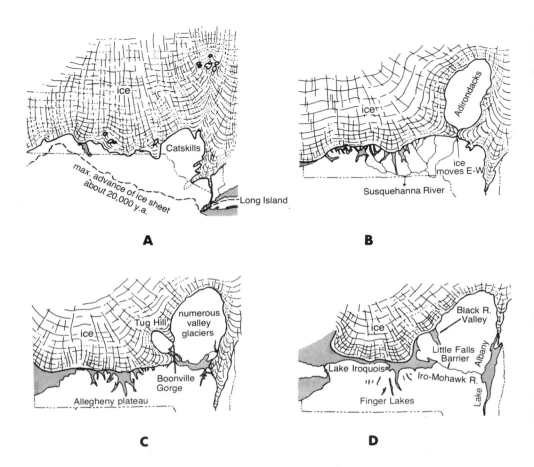

A

ice

Catskills

max. advance of ice sheet about 20,000 y.a.

Long Island

B

ice

Adirondacks

ice moves E-W

Susquehanna River

C

ice

Tug Hill

numerous valley glaciers

Boonville Gorge

Allegheny plateau

D

ice

Black R. Valley

Little Falls Barrier

Lake Iroquois

Iro-Mohawk R.

Finger Lakes

Albany

Lake

Four stages of Wisconsin deglaciation in New York. In B, note that all meltwater is channeled to the Susquehanna River, while ice tongues in the Mohawk Valley move west and east in a kind of "pinzer action" around the Adirondacks. In C, the Finger Lakes are nearly formed, and the Boonville gorge is being cut while valley glaciers are coursing down the sides of the high peaks of the Adirondacks. In D, the Finger Lakes are complete and nearly all Great Lakes drainage goes down the Mohawk River to the Hudson and wears away the Precambrian rock divide at Little Falls. All of the lakes shown in A-C may be termed "proglacial." Courtesy Kendall/Hunt Publishing Co.

Lake Vermont features in particular have been used to chart glacial rebound of the land since the enormous weight of the ice was removed. Tracing originally horizontal shorelines from south to north shows that the St. Lawrence Lowlands surface has risen more than 500 feet. The rebound diminishes south-

Highly polished glacial furrow gouged into the Alexandria Bay gneiss at the entrance to Keewaydin State Park near Alexandria Bay

ward where the ice was thinner and the land less depressed under its weight.

Further recession into Canada eventually left the St. Lawrence Valley free of ice and open to the sea. This led to marine invasion of the St. Lawrence-Champlain Lowlands called the Champlain Sea. The shore and bottom sediments of this embayment may often be recognized by marine fossils they contain. Glacial rebound gradually diminished the inland reach of marine waters to the present condition.

Some of New York's most spectacular gorges and waterfalls are secondary products of the Ice Age. For example, Niagara Falls began when Wisconsin ice backed off the Niagara scarp at Lewiston and subsequent falls migration produced the 7-mile-long Niagara Gorge. The numerous deep, narrow glens, or gulfs, cut into the sides of the Finger Lakes troughs started at the mouths of "hanging valleys" of smaller tributary ice tongues. Many beautiful cascades are found in the northernmost part of the state where streams have cut vigorously downward in response to rebound.

In many places movement of the ice over solid bedrock left scratches called glacial striae which are almost always scraped into polished rock surfaces. These are best preserved in the most durable rocks like the Potsdam sandstone; you won't find many on the limestones or gneisses that have been long exposed to the weather. The scratches result from the grinding of ice-bound rock fragments of all sizes against bedrock. The rocks also grind against each other within the glacier and become similarly marked. Larger boulders may produce broad and sometimes quite deep glacial grooves. At the base of the ice, blocks may press so heavily against bedrock that they cause it to fracture resulting in a string of cuspate joints called chatter marks. The polishing is accomplished by the fine-grained glacial flour that is the abundant by-product of all the grinding action.

Nearly every part of the state contains fields cluttered with boulders of all sizes. These are mostly glacial erratics from somewhere else that the ice carried and then dumped along with other glacial debris. Their prominence at the surface is

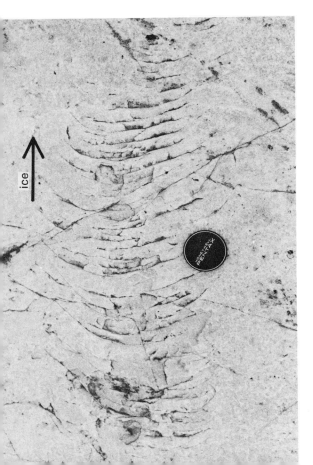

Chatter marks in flat Potsdam sandstone surface at Plessis; arrow indicates ice flow direction.
Courtesy Kendal /Hunt Publishing Co.

often the result of frost-heave by the soil or of tramping of cattle in pastures. Where North Country fields have been cleared for farming the boulders have been used to construct the many stone fences.

Pasture near Lowville, New York cluttered with glacial erratics, messengers from the north dumped here by the ice; some may, in fact, come all the way from Canada. Many North Country fields cleared for farming are bordered with stone fences made from such boulders. Drawing by T. Jancek

View of lower Niagara Gorge from the Power Vista, showing the New Arch Bridge. The strata are well-exposed in the gorge wall. The discharge here is the same as above the falls, owing to release of all the water used in power generation by the Canadian and American plants shown here.

III

THE NIAGARA FRONTIER

The caricature of Niagara Falls on the front of this book was not idly chosen. If New York has a single natural symbol that is almost instantly recognizable to much of the rest of the world, it is Niagara Falls. Niagara Falls are dynamic geology, geology that can be timed on the human clock, geology we can see happening. The account of the formation of the falls and gorge is one of the most fascinating geologic stories ever written. This will not, however, be a comprehensive account. Such is available elsewhere, in my own 1980 book, *Geology Field Guide to Upstate New York* (Kendall/Hunt Publishing Company) and in I. H. Tesmer, Editor, 1981, *Colossal Cataract, The Geologic History of Niagara Falls* (State University of New York Press). The latter book is the most complete and up-to-date account of its kind.

Niagara Frontier is the popular name for the region of the Niagara River. The land is rather flat, but separated into a flight of three "stairs" that ascend from Lake Ontario to the Allegheny Plateau. The lowest "tread" is the Lake Ontario plain floored by late Ordovician Queenston formation redbeds. This is really the lake plain of glacial Lake Iroquois, Ontario's much larger predecessor. The southern edge of the plain terminates abruptly against the steep 250-foot-high Niagara scarp

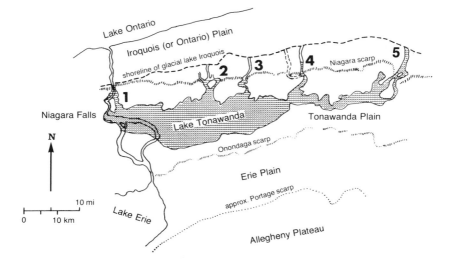

Map of the Niagara Frontier showing plains and scarps. Also shown is the area that was covered by glacial lake Tonawanda, with initial outlets at: 1) Lewiston (mouth of the Niagara Gorge); 2) Lockport; 3) Gasport; 4) Medina; and 5) Holley. The lake is now completely drained and Niagara is the only one of the original falls that remains.

of Silurian strata capped by the late Silurian Lockport group dolostone. The Lockport continues southward as caprock to the second "tread," referred to as the Tonawanda plain, which once held glacial Lake Tonawanda. The Niagara scarp is more than just a local feature, for it extends almost continuously around Lakes Huron and Michigan into northern Illinois. The Tonawanda plain, mostly underlain by weak, late Silurian Salina group beds, extends southward to the region of Buffalo, where it terminates against the rather inconspicuous Onondaga scarp, capped by middle Devonian Onondaga limestone. The upper tread is Erie plain, that includes the Lake Erie basin.

The River

The Niagara River is more like a strait than a river. It is only 33 miles long, and it carries as much water at its head as at its mouth. It has no major tributaries and no mountain origin. It

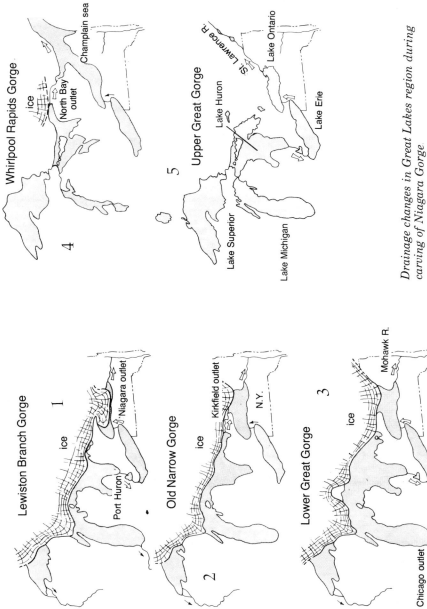

Whirlpool Rapids Gorge

ice

North Bay outlet

Champlain sea

4

Upper Great Gorge

St. Lawrence R.

Lake Huron

Lake Ontario

Lake Erie

5

Lake Superior

Lake Michigan

Lewiston Branch Gorge

1

ice

Niagara outlet

Port Huron

Old Narrow Gorge

ice

Kirkfield outlet

N.Y.

2

Lower Great Gorge

3

ice

Mohawk R.

Chicago outlet

Drainage changes in Great Lakes region during carving of Niagara Gorge.
Courtesy Kendall / Hunt Publishing Co.

has no well-defined valley save for the 7-mile gorge. Incredibly, it carries almost the entire outflow of upper Great Lakes Erie, Huron, Michigan, and Superior, which together with Ontario constitute the greatest freshwater reservoir in the world. Since the lakes are great settling basins, the Niagara River carries almost no sediment.

In making its short land traverse, the river drops 326 feet from Lake Erie to Lake Ontario. Half of that drop is attributable to the falls, and most of the remainder to the rapids just above the falls and in the gorge. The Onondaga scarp is marked by mild rapids near the Peace Bridge at Buffalo.

The Falls

Niagara Falls mark a single great step in the Niagara River. This step has regressed upstream over the past 12,000 years from its origin at the Niagara scarp near Lewiston, New York to its present position 7 miles to the south, leaving the gorge in its wake. The formation of the gorge and maintenance of the falls are inextricably tied together. First, the rock strata now exposed in the gorge walls are almost identical from one end of the gorge to the other. Second, the vertical succession of strata consists of an arrangement of hard and soft rocks that keep the water constantly falling from an overhanging cliff. Third, the rock strata are arranged like a layercake slightly tilted to the south. Thus, the conditions that led to the formation of the falls

View upstream from Spanish Aerocar to Eddy Basin and lower end of Whirlpool Rapids gorge, with key formations labeled
Courtesy Kendall / Hunt Publishing Co.

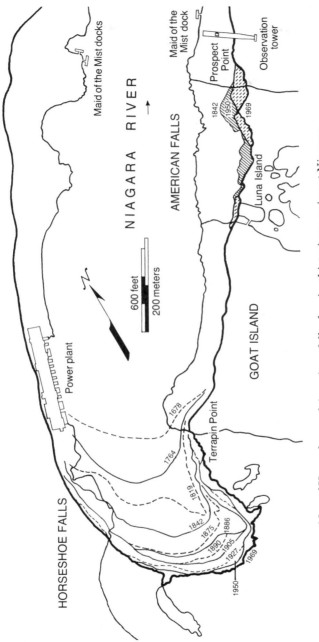

Map of Horseshoe and American falls showing historic recession at Niagara
Courtesy Kendall / Hunt Publishing Co.

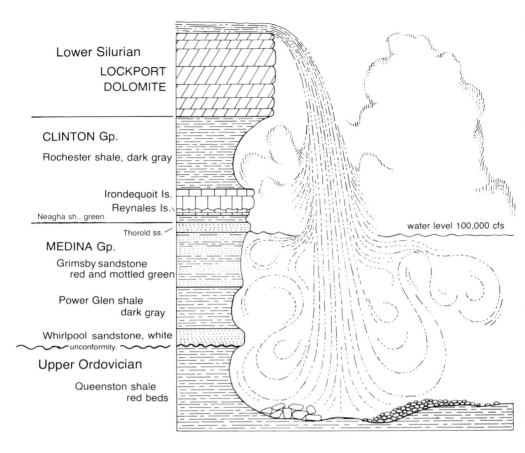

Lower Silurian
LOCKPORT
DOLOMITE

CLINTON Gp.

Rochester shale, dark gray

Irondequoit ls.
Reynales ls.
Neagha sh., green
Thorold ss.
MEDINA Gp.

Grimsby sandstone
red and mottled green

Power Glen shale
dark gray

Whirlpool sandstone, white
unconformity.
Upper Ordovician

Queenston shale
red beds

water level 100,000 cfs

Stratigraphic section under Horseshoe Falls   Courtesy Kendall / Hunt Publishing Co.

in the first place still exist virtually unchanged today. Today, the falls are separated into the Horseshoe Falls on one side of Goat Island, and small Bridal Veil Falls on the other side, which, in turn, are separated from the larger American Falls by tiny Luna Island. The falls have not always been divided this way. Previously there was only one cataract, and Goat Island lay upstream like any of the several islands in the upper river today. The falls divided when the island was intersected and will remain so until the Horseshoe Falls recede beyond the island. By then, the American Falls will be dry.

46

The top of the vertical succession of rock strata, or strati-graphic section, under the falls is an 80-foot thick section of hard dolostone of the late Silurian Lockport group, which is almost immune to the action of water flowing over its top surface. Rocks under the Lockport are mostly soft shales and sandstones of the middle Silurian Clinton and early Silurian Medina groups, and, at the base, the late Ordovician Queenston formation. Thin, resistant beds of Irondequoit and Reynales limestones occur at the base of the Clinton, and Whirlpool sandstone at the base of the Medina. As a whole, this pile of more than 200 feet of generally weak rock presents much less resistance to erosion than the caprock and weathers back more rapidly. The result is that the Lockport is left overhang-ing and eventually breaks off along joint planes and falls into the plunge pool. The whole process is called sapping, and it is responsible for the maintenance of the falls, their upstream recession, and the lengthening of the gorge. In a sense, you can say that the falls have simply moved a section of the Niagara scarp 7 miles south.

The Gorge

Niagara Gorge contains five dimensionally distinct seg-ments, with wide, deep stretches alternating with shallow, narrow ones. They are, from north to south in their order of formation: the wide Lewiston Branch, the Old Narrow Gorge, the Lower Great Gorge, the Whirlpool Rapids Gorge, and the Upper Great Gorge that extends to the falls. The physical character and arrangement of the bedrock strata are nearly uniform from one end of the gorge to the other. Because of this, geologists long ago recognized a direct correlation between the amount of water going over the falls and the size of the canyon being carved underneath. Greater discharge produces a wider, deeper canyon. Lesser discharge erodes a narrower, shallower canyon, generally at a slower rate. The historical recession of Horseshoe and American falls illustrates the point. By an accident of Nature, the Horseshoe Falls have received the lion's share of discharge since the falls split, that now amounts to 90%. As a result, the falls have receded more than one-half mile, while American Falls have moved upstream only a few

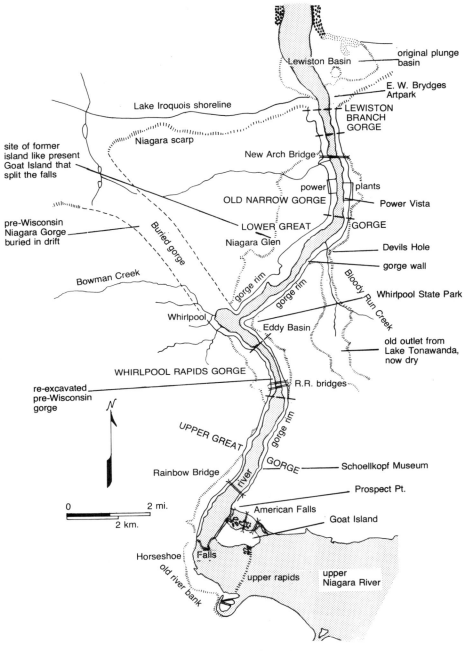

Map of Niagara Falls and gorge

tens of feet. Furthermore, the enormous impact and turbulence of the Horseshoe cataract keeps wearing away the talus and carving a deep plunge basin—deeper, in fact, than the height of the falls—while the American Falls have no plunge basin at all and the lower slopes are perpetually cluttered with huge talus blocks.

Geologists estimate that the Niagara discharge has varied from 15% to 110% of the present flow during the 12,000 years it took to carve the 7-mile long gorge. The explanation lies in the regional drainage changes that accompanied withdrawal of the Wisconsin ice sheet from the Great Lakes region. In general, the retreat of a very thick continental glacier across a landscape of such low relief can be expected to cause profound drainage changes. The land is, at first, depressed under the enormous weight of the ice. As the ice wastes away, an abundance of meltwater collects in every available depression and runs off through every open channelway. The picture changes in complex ways as the ice recedes and opens new outlets. If the whole system is delicately balanced at this stage, glacial rebound may produce further changes—abandonment of old spillways, opening of new ones or old abandoned ones, lowering of lake levels, etc.

The amount of water going over the falls depended on whether, at any stage, the newly-forming Great Lakes Huron, Michigan, and Superior emptied into the Lake Erie basin or found some other bypass outlet, in which case the discharge through Niagara River dropped to low levels.

SITES OF GEOLOGIC AND
SCENIC INTEREST—AMERICAN SIDE

Schoellkopf Geological Museum

For the first-time visitor to Niagara Falls, this small museum located just north of Rainbow Bridge, is a good starting place to get the "lay of the land." Featured there are excellent displays of geology, power development, remedial work in the falls area, rocks, minerals, fossils, a theater program, a geological garden, and more. Several guided tours follow the gorge trails.

The museum stands on the site of the former Schoellkopf Power Station, a major supplier of electricity to the area beginning in 1881. The lower works were destroyed in 1956 when the gorge wall collapsed on top of them. The remains have since been removed. Loss of the facility had a disastrous effect on the economy of the area.

Pre-July 8, 1983 view from Prospect Point to American Falls (nearest) and Horseshoe Falls (farthest), with Terrapin Point profiled
Courtesy Kendall/Hunt Publishing Co.

Similar view on July 9, 1983, with modified profile of Terrapin Point after overhang was blasted away

Prospect Point

From this point, you have a dramatic view of both American and Horseshoe falls, with Terrapin Point profiled in the distance on the far side of Goat Island. Until July 8, 1983, Terrapin Point, with its overlook of Horseshoe Falls, projected farther in an overhanging ledge that showed signs of collapse. Then it was blasted away, and 25,000 tons of rock fell to the talus pile below. Undercutting of the point illustrates the sapping responsible not only for recession of the falls, but also of the sidewalls of the gorge. Before massive hydroelectric development diverted water from the falls, Terrapin Point was under water.

The zigzag crestline of American Falls follows intersecting sets of vertical joints that cut the Lockport caprock. The largest single fall on record occurred on July 28, 1954, when 185,000 tons of rock fell from the rim of American Falls and took some of the viewing area at Prospect Point. This collapse supplied most of the huge talus pile now under the falls. The observation tower at Prospect Point offers a higher view of the falls area. From there, you can also take the elevator down to the river level, to the pathways leading to the edge of the talus pile by the falls, and to the Maid of the Mist boat dock.

Goat Island and Luna Island

Goat Island divides the falls and the rapids above. The rapids descend over steps of the Lockport dolostone that lie stratigraphically above the rimrock. Foot bridges reach Luna Island in the American channel and Three Sisters Islands in the Horseshoe channel. The view of Horseshoe Falls from Terrapin Point is one of the most awesome of all. The falls here plunge 160 to 186 feet over the precipice, depending on flow conditions.

Whirlpool State Park

This park overlooks the giant Whirlpool Basin where the river makes an abrupt right angle turn. The park is about 3 miles north of Prospect Point, and Whirlpool rapids lie in a one-mile stretch of narrow canyon between.

The Whirlpool lies in a re-excavated portion of the buried St. Davids gorge that continues northwestward underground for 3 miles to the Niagara scarp near St. Davids, Ontario. From the outlook at Whirlpool Point, you can see the layercake bedrock strata in the gorge wall at either end of the Whirlpool arc, but none in the debris-filled, old canyon between. The canyon was carved during the pre-Wisconsin interglacial period in much the same way as the present gorge, by recession of falls from the Niagara scarp; but it followed a different course. Then the canyon filled with drift from the Wisconsin ice sheet, and still later the modern Niagara found its present route to the scarp over the Tonawanda plain. When the falls reached the buried gorge, they probably deteriorated to steep rapids for a while as the soft drift sluiced away to hollow out the Whirlpool. A trail leads down to the river where you can walk out on the Whirlpool sandstone shelf at the base of the Silurian section and get a close look up into the Eddy Basin and Whirlpool Rapids. The steep, grass-covered slope in Whirlpool State Park is the bank of the Niagara River formed before the canyon reached this point.

Devils Hole

This is the deep, bowl-shaped plunge basin of the Bloody Run distributary from glacial Lake Tonawanda. It proves that the lake still existed after the main falls had passed this point and discharged some water over the gorge rim. A trail leads down to the river step by step over the rock section of the wall.

The name Bloody Run recalls the massacre of a party of British soldiers in 1763 by the Seneca Indians, who dumped wagons and bodies of men and horses into Devils Hole.

On the opposite side of the canyon, between the Whirlpool and Devils Hole, is Niagara Glen, the site of a former island much like Goat Island today. When the falls reached the island, they split in two. The eastern falls received most of the flow and receded around the island before the other falls had advanced very far, and they were dewatered. The upper channel, called Wintergreen Flats, the plunge basin, and the downstream channel of the western falls are preserved; but the caprock of the island has collapsed to a pile of rubble.

The Power Vista

This is a free visitor area in the Robert Moses Niagara Power Plant. The viewing deck offers magnificent views up and down the canyon and of the Canadian and American power plants. Inside the visitor entryway is a beautiful mural by the late Thomas Hart Benton, depicting Father Louis Hennepin as he first gazed upon Niagara Falls in 1678. The center features various informative displays and a movie.

The power plant access road descends from the south side of the Vista tower to river level past the best-exposed and most accessible stratigraphic section of the entire gorge. Cars are not permitted. The current power development at Niagara was completed under a 1950 treaty between Canada and the United States that, among other things, ensures preservation of the falls. The agreement provides that 100,000 cubic feet per second must flow over the cataracts during the daylight hours in the peak tourist season from April 1 to October 31, or about half of the average discharge of 202,000 cfs. At other times, at night or off-season, the flow may be reduced to 50,000 cfs. The two nations share the excess flow for power generation. In the American system, which is similar to the Canadian, water is diverted from the upper river through two, huge, 4-mile long, underground conduits to the forebay behind the Power Vista. From here, it passes through the power plant at river level to turn 13 turbines and generate electricity. The location of the plant below the rapids section of the gorge enables it to use the maximum drop of the river (305 feet). At night, when electrical demand is low, a portion of the water flowing into the forebay is pumped up to a 1900-acre reservoir through the Lewiston Pump Generating Plant with power supplied from the main station. During peak demand periods, the pumps are reversed to become turbines, and some of the reservoir water is used to generate additional electricity. The water used for power generation is returned to the river, so that the discharge below the plants is equal to that of the upper river.

Earl W. Brydges Artpark

The Artpark is at the mouth of the gorge on the slope of the Niagara scarp behind Lewiston. From there, you have an excellent view of the scarp profile on the Canadian side of the river, and you can look into the fan-shaped plunge basin of the original falls that includes much of Lewiston and part of Queenston across the river. The river widens considerably as it exits the gorge onto the Ontario plain. You may enter the gorge by a haul road in back of the Artpark. The road, built on an old railroad bed, was used during construction of the Robert Moses Power Plant from 1958 to 1963 to haul rock waste from the construction site. A walk along the road offers a view of the complete rock section of the gorge from the Whirlpool sandstone to the Lockport rimrock.

View west from slope above Earl W. Brydges Artpark, with the Niagara Scarp profiled far into Canada
Courtesy Kendall / Hunt Publishing Co.

IV

THE LOWER HUDSON
AND LONG ISLAND

The Palisades Ridge extends about 40 miles from Staten Island northward along the Hudson River to Haverstraw, where it wraps around the northern end of the Newark Basin like a giant sickle. Over most of its length it is a sill, an igneous intrusive body that, as a molten magma, forced its way between the sedimentary strata. The east-west blade of the sickle, however, cuts across the strata and is therefore a dike.

The Palisades Ridge documents a period of volcanic activity of late Triassic time, after the bulk of the redbeds of the Newark Basin had been deposited. This volcanism appears to have started with widespread up-down, or block, faulting of the east coast occasioned by opening of the Atlantic Ocean basin. As the plates pulled apart, tensional forces caused block faulting of the Newark Basin, which lengthened the crust. More extensive dropping of the western border fault tilted the sedimentary strata of the Newark basin to the northwest. It also apparently tapped a deep source of magma that welled up through and then between the strata to form the Palisades sill. The same magma broke the surface farther south to form basalt lava flows in the Watchung Mountains of New Jersey.

The Palisades sill has long interested geologists because it is a good example of differentiation of an originally homogeneous basalt melt into masses of solid rock of different chemical, mineral, and textural makeup. The overall composition of the sill is diabase, essentially the same as volcanic basalt except that it has a coarser texture that reflects slower cooling. Slow cooling and crystallization allowed time for early crystals of heavy olivine to sink and accumulate in a layer of olivine-rich rock. This layer, which is about 10 feet thick and 35-50 feet above the base of the sill, contains about 25% olivine, as opposed to almost none in the rest of the sill. It is easily recognized as a yellowish-brown band of rotten rock that contrasts strikingly with the grayish, solid diabase above and below it.

The sill also varies widely in texture, with the coarsest-grain size in the central part where cooling was slowest, and very fine grain texture in more rapidly cooled "chill zones" near the upper and lower contacts. The chill zones contain blocks of the host sedimentary rock that broke off into the melt. As might be expected, the sedimentary rocks near the sill have been metamorphosed by the heat of the intrusion, producing hard quartzites from original sandstones and fine-grained, unfoliated "hornfels" from shales.

The Hudson Highlands are part of a northeast-trending mountain mass, composed of complexly deformed gneisses and some quartzites and marbles, that extends from Pennsylvania to Connecticut. Numerous linear valleys, especially in the western part of the Highlands, are deeply eroded fault zones and many contain elongate lakes. The gneisses exposed in roadcuts have generally strong layering that dips, or slopes, steeply. These highly fractured and weathered rocks formed in the deep core of the ancestral Adirondack Mountains during the Grenville mountain-building event, some 1100-1300 million years ago. The mountains rimmed the coastline of ancient North America from Labrador to Mexico and rose to great heights. The driving force for that uplift, like that of the much later Acadian mountain-building event, was a collision between Africa and North America.

The Hudson River slices through the Hudson Highlands between Peekskill and Cornwall-on-Hudson in a uniquely beautiful, narrow valley known as the Hudson fjord, that resembles, in many ways, the glacial fjords of Norway. Like its Norwegian counterparts, the Hudson fjord is invaded by Atlantic waters, as it also carries the outflow of fresh water.

Long Island is not much more than a 120-mile long pile of glacial debris; but it's a spectacular one! It is one that geologists have studied extensively because it represents the southern reach of the last, or Wisconsin, glaciation in the New York/New England region. It is a rich museum of end-glacier features, like terminal and recessional moraines, interlobate moraines (between two adjacent ice lobes), outwash deposits, erratic boulders, and knob-and-kettle terrain—the knobs are kames.

The master feature of the Long Island landscape is a plain that slopes gently southward from a maximum of 200 feet at the north shore to sea level at the south shore. The plain consists mainly of outwash sediments with interbeds of till and clay. The surface is capped by two segmented moraine ridges that reach their maximum height of 410 feet above sea level on the western half of the island. The southern ridge, or Ronkonkoma moraine, and the northern, Harbor Hill moraine, form the two flukes at the eastern end of the island with the Great Peconic, Little Peconic, and Gardiners bays between.

The Ronkonkoma moraine records an earlier stand of the ice sheet farther south than the Harbor Hill moraine. Obviously, if the northern were the earlier, it would have been overridden and destroyed by the later advance of the ice. Also, the Harbor Hill moraine apparently was formed after a re-advance of the ice sheet. After having been stabilized at the Ronkonkoma line for a time, the ice front receded to the Connecticut mainland, then re-advanced across the sound to the Harbor Hill stand. The ridges are most widely separated at the flukes; they gradually converge westward and join near Hempstead Bay. Other, lesser, so-called recessional moraines have been recognized in the "Necks" section of the north coast and in the interfluke region. Most of the moraines contain more outwash than till. The bedrock upon which these glacial sediments rest is of Cretaceous age; exposures exist at the western end of Long Island and on Staten Island. Apparently in preglacial time, the island was a ridge connected to the mainland, with a north-facing scarp and gentle south slope; Long Island Sound was the valley of an east-flowing stream. Glaciation scooped out the river valley and dumped the debris on the surface of the island.

Long Island is also an exciting geological laboratory of coastal dynamics where profound changes have occurred since the Wisconsin glacial recession and continue every day. Geologists estimate that the eastern coastline of the island has retreated 10,000-15,000 feet in the last 5,000 years under the attack of waves, wind, tides, currents, and the rise in sea level. Most of the erosion has been on the southeastern shore, which is open to the Atlantic and the prevailing southerly winds; the lion's share takes place during storms.

Long Island is also a treasure trove of early American history, which demonstrates geological influences in human affairs. Thrust like a two-pronged spear between Atlantic and Sound, there is an enormous length of coastline, 600 miles, relative to land area, and an abundance of seafood near at hand. Good well water is plentiful at shallow depths from the fresh water aquifer under the island. The sandy outwash plains are well drained and excellent for certain kinds of farming. Finally, the climate is mild under the moderating influence of the sea; and the eastern part of the island receives more sunshine than any other part of the state. All of these factors contributed to early settlement of Long Island in the 17th century. The western end was first settled by the Dutch, and the eastern part—most of the island—by the English, who migrated from New England. Southold and Southampton were established in 1640, Hempstead in 1643, East Hampton in 1648, Shelter Island in 1652, Oyster Bay and Huntington in 1653, and Sag Harbor in 1660. When the settlers arrived, they found the westernmost part of the island occupied by the Delaware Indians and the rest of the island by the Montauks. The only current reminders of the Indian heritage are names like Shinnecock, Manhasset, Massapequa, and, of course, Montauk.

Palisades Interstate Parkway
38 mi./62 km.

The Palisades Interstate Parkway provides a rapid, 38-mile link between the George Washington and Bear Mountain bridges. Over much of its length it is like a ribbon of parkland, unusual for a highway so near the metropolitan region. The southernmost 11 miles is entirely within New Jersey where the road gets its name, for it is there that it flirts with the cliff edge of the New Jersey Palisades. The central portion traverses inland from the Hudson River across the Triassic redbeds of the Newark basin in Rockland County. The northern, faulted limit of the basin is crossed near West Haverstraw, and the remaining 10 miles of parkway winds through hilly, wooded country of the Hudson Highlands, finally skirting the north slope of Bear Mountain to Bear Mountain Bridge.

Most of the scenic value of the Palisades accrues to Manhattan, Bronx, and a strip of Westchester County bordering the Hudson, including Yonkers. Across the river from uptown Manhattan, the imposing cliff rises a full 400 feet above the water. The level increases northward to more than 700 feet at Hook Mountain near Nyack and ultimately to 827 feet at High Tor which rises precipitously above the river's edge at Haverstraw.

Altogether about 1000 feet thick, only 200-300 feet of the lower part of the sill are exposed in the Palisades cliff, while contact with the underlying shales is concealed under a steep, tree-shrouded talus slope made of rocks fallen from the cliff. Disintegration of the rock depends upon an abundance of open fractures called joints, into which

Storm King Mt., overlooking the north entrance of the Hudson fjord; Catskill Aqueduct passes under the river here

strategic choke point during Revolutionary War

Cornwall-on-Hudson

Hudson Highlands

Storm King Mt.

West Point
218

Hudson fjord, narrow slot through the Highlands gouged by an ice tongue during the Ice Age

rugged Hudson Highlands terrane developed on resistant Precambrian metamorphic rocks

Highland Falls

folded sedimentary rocks

gneisses

Bear Mt.

Peekskill granite

Bear Mt., with excellent views of the Hudson fjord and Bear Mt. bridge from the highway to the summit

Jones Pt.

Cortlandt complex

Bear Mt. Inn, located on the bedrock floor of the pre-glacial Hudson; sites of Forts Clinton and Montgomery; a huge chain was strung across the river here during Revolutionary War to prevent invasion of British warships

crossing Ramapo-Canopus boundary fault between Triassic Basin and Highlands; greater resistance of Precambrian rocks revealed by higher elevation

Stony Pt.

Haverstraw

landing site of Sir Henry Clinton and troops before taking American forts at Bear Mt. on October 6, 1777

crossing the "point of the sickle" where the sill has become a cross-cutting dike that juts from the redbeds

9w

Yonkers-Fordham gneisses

Manhattan prong

87

Palisades Parkway

Nyack

Tappan Zee

N.Y.

huge quarry in Palisades sill

Sparkill

State Line lookout, 532' above river, good views of columnar jointing in cliff face

Palisades

crossing Sparkill Gap, a pre-glacial Hudson River channel

Alpine lookout, good profile views of Palisades cliffs north and south, polygonal joint (fracture) pattern visible in bedrock surface of lookout

Triassic red beds

HUDSON RIVER

Inwood marble

N

0   10 mi

0   10 km

Spuyten Duyvil Creek, and its south extension, Harlem River, are carved in a band of weak Inwood marble

Harlem R.

George Washington Bridge

Manhattan

Palisades sill

Rockefeller lookout, with views of George Washington Bridge and Manhattan, with Spuyten Duyvil at its north end; lookout is 400' above river

PALISADES INTERSTATE PARKWAY

60

Geologic Era	Millions of Years ago	Units shown on route map
MESOZOIC	195	**Newark Basin** Triassic Newark group red conglomerate, sandstone, shale Palisades diabase
PALEOZOIC	350	**Manhattan Prong** Devonian Peekskill granite
	435	Ordovician Cortlandt & Croton Falls complexes
		Cambrian-Ordovician Inwood marble Manhattan schist Lowerre quartzite
pC PROTEROZOIC	575-1100	Yonkers granite gneiss Fordham gneiss
	1100-1300	Hudson Highlands gneisses, some marbles & quartzite

Times of formation of various rocks in southeast New York. All but the Triassic Newark group rocks were formed as North America and Africa converged; the Newark Basin dropped in response to opening of the modern Atlantic Ocean basin. Symbols are the same as on the accompanying map.
Adapted from Isachsen (1980).

water seeps and freezes, or plant roots work their way, prying the rock apart. The most conspicuous joints are vertical, dividing the rock face into polygonal columns that resemble stakes set firmly in the ground as in a defense enclosure; hence the name, palisade. Joints like these are a trademark of volcanic lava flows and sills, resulting from contraction that accompanies solidification of the melt. A less prominent set of horizontal joints aids piecemeal destruction that locally forms a stair-step arrangement of columns of different heights.

Three overlooks on the New Jersey section of the parkway provide excellent views across the river. The Rockefeller Lookout, three miles north of the George Washington Bridge, is the best place to see the bridge and the Manhattan and Bronx skyline. Spuyten Duyvil Creek, between the northern end of Manhattan and the Bronx, lies almost directly across the river. The creek continues as the Harlem River separating the northeast side of Manhattan from the Bronx, and then flows south into the East River. The stream courses lie in a band of weak Inwood marble of the Manhattan Prong, which is named for good exposures in Inwood Hill Park, visible in the high ground just south of Spuyten Duyvil.

Alpine Lookout is three miles north of Rockefeller Lookout and four miles south of State Line Lookout. The hilly country across the river from the northern lookouts reflects the high resistance to erosion of the underlying Fordham gneiss in the Bronx, Yonkers, and Hastings-on-Hudson. The State Line Lookout offers the best profile views of the cliff-and-talus slopes of the Palisades to north and south. Each lookout permits close examination of the sill, and a good view of the polygonal pattern of columnar jointing exposed at the top of the cliff.

Palisades cliff and Hudson River from State Line lookout, with strong columnar jointing

Between State Line Lookout and Thiells (18 miles), the parkway traverses inland across fairly level redbed country of the Newark basin. Three miles north of the lookout, it passes over marshy Spar-kill Creek which, just to the east, has cut a deep slot into the scarp at the Hudson River's edge. This is also on the north side of a 2-mile wide saddle in the Palisades Ridge called the Sparkill Gap, that apparently marks the course of the preglacial Hudson River.

The tapered end of the Palisades dike at the "point of the sickle" crosses the road 2 miles south of Thiells, near West Haverstraw, forming a sharp ridge that rises 400 feet above the east side of the highway. Hills to the west are the Ramapo Mountains of the Hudson Highlands, on the upthrown side of the Thiells fault, bordering the Newark basin.

Between Thiells and Bear Mountain Bridge (10 miles), the road winds through wooded and lake-dotted hills of the Hudson Highlands in Harriman State Park.

The summit of Bear Mountain is the best place for a sweeping view of the Hudson Highlands landscape. It is easily reached by the Perkins Memorial Drive, one of three mountaintop roads in New York; the other two are on Whiteface Mountain in the Adirondacks and Prospect Mountain by Lake George. From there you also have a spectacular view of the Bear Mountain Bridge and the Hudson fjord, that narrow, steep-sided, zig-zagging stretch of valley carved through the Highlands first by the river, then by glacial ice, and now re-occupied by the river and swept by Atlantic tides every day. The summit is almost bare, made up of pinkish, massive Storm King granitic gneiss, like that of Storm King Mountain north of West Point, nine miles farther upriver.

See map page 60.
US 9W, NY 218:
Sparkill Gap—West Point—
Storm King Mountain
34 mi./55 km.

Sparkill Gap is a two-mile wide saddle in the Palisades Ridge that marks the abandoned course of the preglacial Hudson River. The present course is different because of Ice Age glaciation. Between Sparkill and Nyack (4 miles), US 9W descends the talus slope of the Nyack Range of the Palisades Ridge, with excellent views of the extremely broad section of the Hudson estuary called the Tappan Zee, and of the Interstate 90 Tappan Zee Bridge. The broadness of this section of the river valley reflects the underlying soft shales, sandstones, and arkoses of the Triassic Newark group, which were easily worn away by river and glacial erosion. South of Sparkill, the river is more narrowly confined between the harder rocks of the Palisades sill and the Manhattan Prong. On the north, between Peekskill and Cornwall-on-Hudson, is the narrowest segment of the lower Hudson through even harder rocks of the Hudson Highlands.

Between Nyack and Haverstraw (9 miles), the road climbs back up onto the Hook Mountain section of the ridge where it begins a long arch to the west as the interlayered sill becomes a dike that cuts across the bedding of the Newark redbeds. About two miles south of Haverstraw, the road descends across the face of the cap past an enormous quarry in the Palisades diabase. The density and hardness of this rock invited the interest of quarrymen as far back as the post-Civil War period, when it was used to make cobblestone pavements in the streets of New York and other cities. Around 1890, New York City began extensive hard surfacing, or macadamizing, of its streets with crushed stone and asphalt. The Palisades rock was found to be ideal for this purpose, and for railroad ballast. Over the next few years, several quarries were begun, some operating around the clock, blasting day and night, and destroying the scenic cliff. Public outcry and the election of conservation-minded Teddy Roosevelt as governor of New York in 1899 brought the destruction to a virtual halt with the establishment of the Palisades Interstate Park in 1909.

The Palisades sill and dike belong to a class of shallow, igneous bodies formed in volcanic environments where ascending magma did not quite make it to the surface. A trademark of such bodies is much fracturing, or jointing, that results from contraction during cooling and solidification of the melt. It is this jointing, so visible in the Palisades cliff, as well as the high density, uniformity, and durability, that makes the diabase valuable for certain types of quarry operations. The rock can easily be broken up by blasting into large angular blocks ideal for riprap in revetments or jetties along the seacoast. Crushed to gravel size, the rock makes ballast for railroads, underpavement for highways, and aggregate for asphalt and concrete.

Between Haverstraw and Stony Point (5 miles), the road traverses the northernmost redbed lowlands of the Newark basin near the Hudson River. The bedrock of Stony Point itself, a knobby promontory in the Hudson, is part of the late Ordovician Cortlandt igneous complex. This is a unique geologic feature consisting of several intrusive bodies that contain a wide range of dark rocks, some quite rare. The bulk of the complex lies across the river south of Peekskill, the site of the only commercial emery deposits in the United States. The emery, used primarily as an abrasive, formed by metamorphism and chemical alteration of the Manhattan schist adjacent to the intrusion.

Between Stony Point and Jones Point (5 miles), the route follows a northeast-trending arm of the river excavated in the Thiells fault zone, one of the border faults of the Newark basin. Dunderberg Mountain, the backbone of Jones Point, is on the upthrown, or Hudson Highlands side of the fault and is held up by granitic gneiss similar to that of Bear Mountain and Storm King Mountain. This is the southern gateway to the Hudson fjord through the Highlands, where the river makes a sharp 100 degree bend to the northwest.

Views of Bear Mountain Bridge and the Highlands Gorge on the southern approach from Jones Point are among New York's loveliest scenes. The bridge itself is an elegant and simple structure that contrasts strikingly with the steep, wooded slopes of the mountains at either end of it, Bear Mountain on the west and Anthonys Nose on the east.

Bear Mountain Bridge between Anthony's Nose (left) and Bear Mountain (right of picture)

Bear Mountain Bridge spans one of the two choke points of the Hudson Gorge; the other is at West Point. During the Revolutionary War, the patriots saw this place as a key point of defense against British invasion of upstate by way of the Hudson and built two forts at the base of Bear Mountain: Forts Montgomery and Clinton. Also, a log boom and chain were strung across the river along the approximate line of the present bridge. All of this was to no avail, for on October 6, 1777, both forts were taken from behind by British troops under Sir Henry Clinton. They had landed at Stony Point and then marched around Bear Mountain along the route of the present Palisades Parkway. The British dismantled the chain, allowing warships to sail on to Kingston, then the capital of New York, and burn it to the ground. Later, with the Hudson Gorge again under American control, West Point was seen as a better choke point, and similar defenses were built there.

The broad, flat shelf at the base of Bear Mountain where the forts were built, and where the famous Bear Mountain Inn now stands, is the bedrock floor of the preglacial Hudson, here about 160 feet above the modern river. Similar erosional remnants of this old channel exist on both sides of the river over much of the length of the gorge, all at about the same elevation. The largest and most conspicuous is the site of the U.S. Military Academy. The platform at Bear Mountain was carved in gneisses bordering the Storm King granite gneiss; banana-shaped Hessian Lake wraps around the base of the mountain and occupies a basin scooped from the contact zone by glacial ice.

Hudson fjord from Trophy Point; Storm King is farthest promontory on left.

Between Bear Mountain Bridge and the Storm King Highway juncton (6 miles), US 9W traverses inland, looping around the city of Highland Falls and the military academy.

Certainly the most spectacular view of the Hudson fjord is at Trophy Point at the northern edge of the so-called "plain," or parade grounds, of the Academy. The river makes a peculiar S-shaped bend here that, more than any other physical attribute, makes it such an ideal site for defense of the river. The second log boom and chain were strung here in 1778 after the siege of Kingston. At the same time, Fort Putnam was built on the high rock shoulder, called Crown Hill, behind the plain, with excellent visibility over the river. These formidable defenses were never tested by the British before their surrender at Yorktown, Virginia, in 1781.

The S-bend in the river at West Point is, perhaps, the most profound glacial change in the gorge. Tertiary, or preglacial, erosion by the river had whittled a wide bedrock channel with a more moderate bend, now preserved in the flat shelves on both sides of the river and on Constitution Island.

Between West Point and Storm King Mountain (3 miles), the Storm King Highway is notched into the steep, wooded, glacially steepened gorge wall underlain by the tough Storm King granite gneiss. Just south of the mountain, the road descends to cross The Clover, carved in the crush zone of a northeast-trending fault. The fault extends across the river to Breakneck Brook along the southeast-facing scarp of Breakneck Ridge. The circuit around the granite rib of Storm King is another scenic high point of the Hudson Gorge. The road affords views south to West Point and north to the open water beyond the Hudson Highlands. Here one can best appreciate the enormous pressures involved when some of the southward-advancing ice sheet jammed into the gorge. The depth of glacial gouging was not known until 1909, when test borings made for the Catskill aqueduct found the bedrock valley to be nearly 1000 feet below water level, or 1160 feet below the preglacial channel! The present channel is only about 80 feet deep, so the sediment fill, in places, exceeds 900 feet. The aqueduct tunnels through the Storm King gneiss under the river from a point north of the mountain to the end of Breakneck Ridge, where there are two pumping stations.

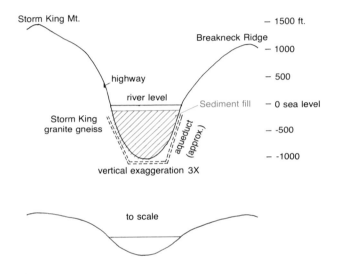

Cross-section of Hudson fjord at Storm King

NY 25A, NY 27, NY 25:
Queens—Orient Point
95 mi./154 km.

NORTH SHORE OF LONG ISLAND

Between the junction with Interstate 295 in Queens and Northport, NY 25A skirts the "Necks" section of Long Island's north coast on the north slope of the Harbor Hill moraine. The Necks are interbay divides that separate Little Neck Bay, Manhasset Bay, Hempstead Bay, Cold Spring Harbor, Northport Harbor, and others. The bays appear to represent the courses of short, north-flowing preglacial tributaries to an east-flowing river at a time when the island was part of the New England coast. The bays display the long, pointed form typical of drowned river mouths everywhere. Erosion and sedimentation of the necks coast since glacial recession and filling of the sound with Atlantic waters has been minimal because of its sheltered position at the narrow west end of the sound. Erosional modification increases eastward, resulting in a smoother coastline, blunted, wave-eroded necks, and development of beaches and spits.

Two north-south-trending interlobate moraines, called Manetto Hills and Half Hollow Hills, are south of Huntington. North of Huntington on Lloyd Neck is Caumsett State Park, where one has access to the pebbled beach and can examine the glacial materials of the seacliff. This section of coast is constructed of the so-called Sands Point recessional moraine, "recessional" because it was formed where the ice stood after receding northward, rather than advancing and then stabilizing. The steep cliffs display badlands erosion, with many closely spaced angular ravines with sharp ridges between them. They are made up of well-bedded pebble gravels with lots of white quartzite, some sand lenses and layers, and occasional clay. Strewn

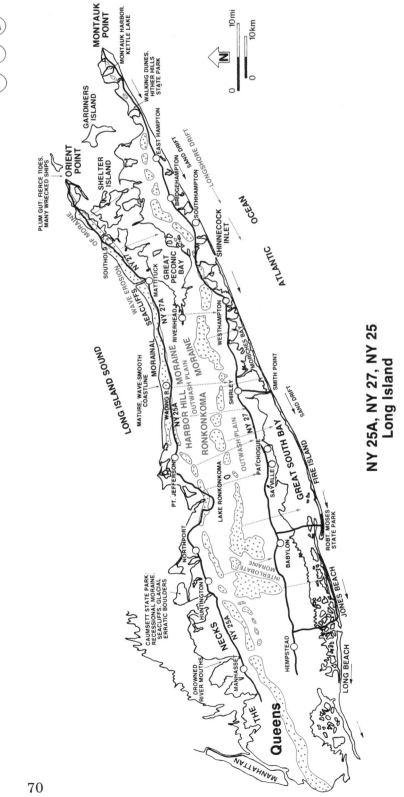

**NY 25A, NY 27, NY 25
Long Island**

CAUMSETT STATE PARK:
RECESSIONAL MORAINE.
SEACLIFFS, GLACIAL
ERRATIC BOULDERS

DROWNED
RIVER MOUTHS

THE
NECKS

NY 25A

MANHASSET

HEMPSTEAD

HUNTINGTON

NORTHPORT

PT. JEFFERSON

LAKE RONKONKOMA

BABYLON

INTERLOBATE
MORAINE

ROBT. MOSES
STATE PARK

JONES BEACH

LONG BEACH

GREAT SOUTH BAY

FIRE ISLAND

SAYVILLE

PATCHOGUE

NY 27

OUTWASH PLAIN

RONKONKOMA

HARBOR HILL

Queens

MANHATTAN

LONG ISLAND SOUND

MATURE, WAVE-SMOOTH
COASTLINE

MORAINAL
SEACLIFFS

WADING R.

NY 25A

MORAINE

OUTWASH PLAIN

MORAINE

SHIRLEY

SMITH POINT

SAND DRIFT

MORICHES BAY

MORICHES BAY

WESTHAMPTON

RIVERHEAD

NY 27A

MATTITUCK

GREAT
PECONIC
BAY

WAVE EROSION

SOUTHOLD

OF MORAINE

SHELTER
ISLAND

PLUM GUT: FIERCE TIDES.
MANY WRECKED SHIPS

ORIENT
POINT

GARDINERS
ISLAND

SHINNECOCK
INLET

SOUTHHAMPTON

BRIDGEHAMPTON

SAND DRIFT

EAST HAMPTON

LONGSHORE DRIFT

ATLANTIC

OCEAN

WALKING DUNES.
HITHER HILLS
STATE PARK

MONTAUK HARBOR.
KETTLE LAKE

MONTAUK
POINT

N

0   10mi

0   10km

70

Pebbles from north Long Island beach near Southold, with a lot of quartzite, some conglomerate, and a variety of metamorphic rock types.

about the gravelly beach and offshore are large boulders representing a wide variety of metamorphic rock types, igneous basalts, red conglomerate, and others. These are glacial erratics, "messengers" from the mainland to the north that were brought here by the ice, dumped, and only recently liberated from the moraine.

The Harbor Hill moraine converges with the shore at Port Jefferson. Port Jefferson Harbor, another drowned valley, differs markedly from the neck bays in the smooth wave truncation of the headlands and the construction of a bay mouth bar by longshore drift of sand. The same is true in the small Mt. Sinai Harbor two miles to the west.

Between Mt. Sinai Harbor and Orient Point, the Harbor Hill moraine is hard against the shore, and there are no bays at all; the shoreline is smooth, cuspate and unbroken, in striking contrast to the necks coast. For the most part, the sea cliff is steep and high, exposing the interior of the moraine, which is made up of the same basic

71

North shore beach strewn with glacial erratics liberated from the Harbor Hill moraine by wave erosion   Drawing by T. Jancek

materials as those of Caumsett State Park. Unfortunately access to, and use of, all publicly-owned north-shore beaches is restricted to residents of the various towns, who must have a permit.

Between Wading River and Orient Point, we follow the northernmost highway, including segments of NY 27 and 25. Over nearly this whole distance, the moraine ridge is visible just to the north, and the road traverses the outwash plain pocked with numerous kettles of widely varying sizes. Access to the north beaches is by way of infrequent roads that pass through natural slots in the moraine; in many places, one must descend by foot over the last steep part.

Between Mattituck and Orient Point, the Harbor Hill ridge has been virtually eliminated by water erosion, while the pitted outwash plain descends to sea level. The Point itself, on private land, is a mere fingertip of sand that contrasts markedly with the bold headland of Montauk Point 20 miles to the southeast across Gardiners Bay and Block Island. The moraine emerges again on Plum Island, a mile and a half offshore to the east. The narrow strait between is the legendary Plum Gut, through which fierce tides each day flush and fill Long Island Sound with Atlantic waters.

Montauk Point; rapid wave erosion of the Ronkonkoma moraine now threatens to destroy the lighthouse.

See map page 70.
NY 27:
Montauk Point—Queens
115 mi./186 km.

Montauk Point is one of the windiest places on the Atlantic coast. Thrust into the Atlantic 20 miles farther east than Orient Point, it is the easternmost tip of New York State, in constant battle with the sea. The Ronkonkoma moraine extends right to the point and presents a bold promontory of glacial till. Nevertheless, the fierce tides of the "Rip" between the point and Block Island, and the unending onslaught of wind and wave, especially during the frequent winter storms, have destroyed much of the headland, which loses about 10 feet each year. The Montauk Light is now only 50 feet or so from the brink of the seacliff. When the original lighthouse was built on the same spot in 1795-97, it was about 300 feet from the edge! The original structure, incidentally, had the distinction of being the first lighthouse to be federally financed by the newly-formed United States of America. The Army Corps of Engineers recently estimated that two-thirds of the sediment supply for all the southern beaches and barrier islands comes from erosion of Montauk Point and the balance from onshore drift of outwash sediments from the shallow, gently sloping sea floor south of the island.

The Montauk till is grossly different from the sediment of the Harbor Hill moraine along the north shore. This is a true "boulder clay," a conglomeration of materials of all sizes indiscriminately ferried from the north by the glacier, and then unceremoniously melted out and dumped in place at the ice margin. An excess of clay binds the whole mass together. Erosion, especially by rainwash, landslides, waves, and wind, sorts these materials out, carries the fine particles away, and leaves the beach littered with glacial erratic boulders and gravel. Reworked over and over by the waves, the coarse fraction is gradually worn down to sand that drifts westward to supply the beaches all the way to Jones Beach, nearly 100 miles away.

As in the case of the Harbor Hill moraine on the north fluke, the Ronkonkoma moraine originally continued eastward from Montauk. Remnants remain today on Block Island, 15 miles out across the "Rip," and on Martha's Vineyard and Nantucket Island off the Massachusetts coast.

The 11-mile section of NY 27 between Montauk Point State Park and Hither Hills State Park, the Montauk Point State Boulevard, is one of the most scenic drives in New York state, a magnificent blend of beach and dune seascape to the south, and knob and kettle moraine to the north. The route skirts the south shores of three exceptionally large salt water kettle lakes; the easternmost Oyster Pond, and the westernmost, Fort Pond, are both a little less than a mile in maximum breadth, while Lake Montauk in the middle, with a length of over 2 miles, is the largest kettle lake on Long Island.

At the western end of this beautiful drive are the intriguing Walking Dunes of Hither Hills State Park, where blowing sand has covered over and heightened the hummocky moraine topography. As the dunes migrate or "walk," they engulf the scrub oaks that grow there and slowly strangle them. Later, when the sand moves on, the trees emerge, leaving an eerie scene of ragged, blackened trunks jutting from white sand.

This outer segment of the south fluke was once an island, separated by a gap adjacent to Hither Hills. Napeaque Beach now bridges that

Hither Hills "walking dunes," showing scrub oak stumps that were earlier buried by sand migration

gap, having been formed by sand drifting westward from Montauk Point. This accounts for the low, sandy, and marshy strip between Hither Hills and East Hampton.

Between East Hampton and Southhampton, the highway skirts several small bays, including Hook Pond, Georgica Pond, Wainscott Pond, Fairfield Pond, Sagaponack Pond, Mecox Bay, Channel Pond, Jule Pond, Sayre Pond, Phillips Pond, Wickapogue Pond, Agawan Pond, Cooper Neck Pond, and Halsey Neck Pond. With the single exception of Sagapanock Pond, which has a narrow tidal inlet, every one of these bays is now completely closed off by baymouth bars formed by westward drift of beach sand. The companion wave erosion has truncated all of the headlands to produce an exceptionally smooth coastline. The bays originated as drowned river mouths, estuaries. One to three miles north of the road, the Ronkonkoma moraine forms a segmented ridge that parallels the north shoreline of the fluke near Nyack, Little Peconic, and Great Peconic bays.

Southampton marks the eastern end of the Long Island's barrier islands or beaches, which extend all the way to Jamaica Bay and protect the deeply indented estuarine coast from the open sea. The islands include the Shinnecock Bay barrier, or Southampton and Tiana beaches, Hampton Beach, Pikes Beach, Great South Beach or Fire Island, Gilgo Beach, Jones Beach, Long Beach, and Rockaway Beach. These narrow, offshore strips of sand react with extreme sensitivity to the wind, waves, and longshore currents, and constantly change. They exist on the south shore and not on the north for two principal reasons: 1) the outwash plain descends very gradually below sea level so that the water is shallow and there is a plentiful supply of sediment; 2) the shore is open to the Atlantic with strong prevailing southerly winds, more intense wave erosion, and stronger currents.

Separation of the barrier islands from the land increases from east to west, reaching a maximum of about 6 miles at Jamaica Bay. Over that long distance, the islands are breached by only 4 tidal inlets, at Shinnecock and Moriches bays and the western ends of Fire Island and Jones Beach. The broad lagoons behind the barrier islands protect the coast from direct onslaught of the sea; as a result the south coast has seen extensive development, urban sprawl, at the western end.

On the south shore of Long Island between Southampton and Queens, NY 27 traverses the level outwash plain south of the Ronkonkoma moraine, skirting numerous estuarine bays that open into the lagoon. These estuaries are actually meltwater channels formed at the height of Wisconsin glaciation, when the ice margin stood just to the north and the region to the south was dry land. The road also passes through numerous historic villages established long before the American Revolution. It is interesting to visit the cemeteries in these towns and to read the still-intact inscriptions on the old tombstones. Many, especially those of marble or soft sandstone, are no longer readable because these materials weather so easily; there are, however, many from the 1600s that are made of more resistant slate, and on which the inscriptions are still clear.

FIRE ISLAND NATIONAL SEASHORE

Fire Island is a beautiful place as well as a museum of dynamic geology where drastic changes may occur "in the blink of an eye." It is also an outstanding example of the often catastrophic effects of human misunderstanding of and interference with nature.

Fire Island is the longest of the south shore barrier islands, stretching 32 miles from Moriches Inlet on the east to Democrat Point on the west, unbroken by any other inlet, at least at this writing. The eastern end is separated from the mainland by Moriches and Narrow bays, while the western three-quarters is separated by the broad expanse of lagoon called Great South Bay. In 1980, part of the island was designated by Congress as wilderness, the only area of New York so honored, and Fire Island National Seashore was established. Most of the island within the authorized seashore boundaries is still privately owned. Much of the balance is as wild today as it was 400 years ago when the first Europeans saw it, and the National Park Service is striving to keep it that way. Car access to the seashore is limited to the extreme eastern end at Smith Point, and by way of a causeway from West Islip to Robert Moses State Park at the extreme western end. Ferries also operate from Patchogue and Sayville between May and November, carrying visitors to the Watch Hill and Sailors Haven sections of the seashore, respectively. No matter how you get there, the only way to see the natural wonders of the island is on foot.

Fire Island is one of 282 barrier islands strung, like linked sausages, along the American coastline from Long Island to Brownsville, Texas. On a worldwide scale, barrier islands are almost exclusively a feature of gently sloping coasts. All such coasts are sedimentary, and the sediments are easily reworked by waves and currents in the shallow water.

All the barrier islands on the southeast of Long Island probably began as beach ridges and dunes that lined an earlier coastline. As sea level rose, these became linear offshore islands and the low areas

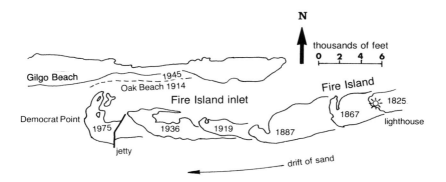

Progressive growth of Democrat Point as a result of longshore drift of sand

behind became brackish lagoons. The process has been aided and, in fact, the islands have been largely perpetuated by the westward drift of sediments generated by rapid erosion of the eastern end of the island. In 1825, the Fire Island Lighthouse was at the western tip of the island. Westward drifting sand has since extended the island as a sand spit so its present terminus, Democrat Point, is nearly 5 miles west of the lighthouse; and overlaps the eastern end of the Jones Beach barrier island. The growth rate now exceeds 200 feet per year!

The stability of a barrier island depends on several factors, of which a plentiful supply of sand is the most crucial. In addition to the influx of sand from eastern Long Island, Fire Island also receives sediment from waves advancing inland over the shallow bottom. Storm tossed waves often wash over parts of the island depositing more sand at higher levels. Winds pile the sand even higher in dunes behind the beach. Storms often also wash a slot through the island to form a new inlet, which may survive for a long time, or may soon be closed by sedimentation. The Fire Island inlet at the western end is now artificially maintained by dredging every 2-3 years.

Fire Island and the other barrier islands of the south shore gradually migrate inland as the sea level rises. Meanwhile, dune and overwash deposits override lagoon deposits, with the combined effect of shifting the whole assemblage landward. Sea level is still rising, and the current rate of shoreface retreat is about 2-4 feet per year. Summer homes have been built virtually at the water's edge with little regard for this natural threat, and millions of dollars worth of property damage has been inflicted by storms, when most of the erosion takes place. Development, including planting, also stabilizes the dunes, preventing further landward migration of the island, while the shoreface is continually eaten away and the beach gets narrower. The ongoing rise in sea level without landward migration of the island also tends to diminish the offshore supply of sediment, so that beach replenishment is increasingly restricted to the longshore drift from eastern Long Island. The beachfront results of these processes are easy to see; the beach is narrow, a sea cliff has been carved into the dune front, and many of the existing houses perch precariously on the brink of the cliff. How long can this go on?

Seacliffs cut in dunes of Fire Island

V

THE NEW ENGLAND FRONT

The Hudson River Valley slices through the middle of this section from Glens Falls to New York City. It continues seaward as a submarine canyon of immense proportions for well over 100 miles from lower New York Bay to the edge of the continental shelf. During the height of Wisconsin glaciation, so much water was locked up in the continental ice sheets that sea level was lowered 330 feet. Much of what is now submerged continental shelf was then dry coastal plain. As the ice sheet later retreated, the Hudson, then enormously swollen with silt-laden meltwaters, was able to carve its deep bedrock canyon.

The Manhattan Prong is a geologic subprovince of southeastern New York that encompasses Manhattan, the Bronx, most of Westchester County, and a corner of Putnam County. Most of the rocks are metamorphic, including the Grenville age (1100 million years old) Yonkers and Fordham gneisses and the Cambrian-Ordovician (about 500 million years old) Manhattan schist, Inwood marble and Lowerre quartzite. Near Peekskill, adjacent to the Hudson Highlands, these rocks are intruded by the Ordovician (435 million years old) Cortlandt and Croton Falls complexes and the Devonian (350 million years old) Peekskill granite. The landscape is one of rolling hills held up by the harder gneisses and schists and valleys carved in the weaker marble.

The Taconic Mountain Range is composed of complexly deformed schists, phyllites, slates, marbles, gneisses, and quartzites belonging to the structural unit called the Taconic Klippe. A klippe is a mass of rock that has been moved bodily over other rocks to a new location, either by thrust faulting or tectonic landsliding. The Taconic Klippe is very large, stretching in a narrow band for 150 miles from its boundary with the Hudson Highlands near Patterson to Ticonderoga. Related Taconian thrust faulting continues farther north to the Canadian border. The klippe is riven with numerous slides, faults, and associated folds and crush zones, and innumerable thrust fault slices are stacked on top of each other. The Taconic Klippe, therefore, is really a composite of many structural elements, including many smaller klippen, as well as erosional "windows" through some of the fault slices. The complex mass originated in the region to the east and moved to its present position during the Taconian mountain-building event.

In Dutchess County, the grade, or temperature, of metamorphism, indicated by certain minerals in the rocks, increases rapidly from west to east, going from low-grade slates to high-grade sillimanite gneiss in only a few miles. Sillimanite is a metamorphic mineral typically crystallized near the melting temperature of the rock. Farther north, low-grade slates and phyllites dominate the rocks of the range, and higher grades of metamorphism are not represented.

The Shawangunk conglomerate rests on the Taconic unconformity, a surface developed by long erosion of the underlying Ordovician bedrock. The sediment came from wearing down of the ancestral Taconic Mountains to the east. The clean sand and quartz pebbles are deposits reworked by wave action along the shore of a sea that lapped against the west side of the mountains during Shawangunk time. Many of the quartz pebbles are shaped almost like dishes, a form that tells of prolonged washing in the surf.

Another unconformity, exposed in several places north of Albany, is the contact between the Grenville-age rocks of the Adirondacks and the Cambrian Potsdam sandstone. This astounding gap of nearly 600 million years in the rock record represents a period of Earth history during which as much as 15 miles of rocks were eroded from the ancestral Adirondacks. The end result was a landscape of low relief. The Potsdam Sea then invaded the region to initiate Paleozoic sedimentation and a new rock record.

Lakes George and Champlain are both post-Taconian fault depressions, or graben, gouged by ice during the Ice Age. During Wisconsin deglaciation, while the northern part of the Champlain Valley and the St. Lawrence Valley were still blocked by ice, a proglacial lake called Lake Vermont filled the depression and extended far beyond the present lake shores. Still later, when the St. Lawrence opened up, Atlantic waters

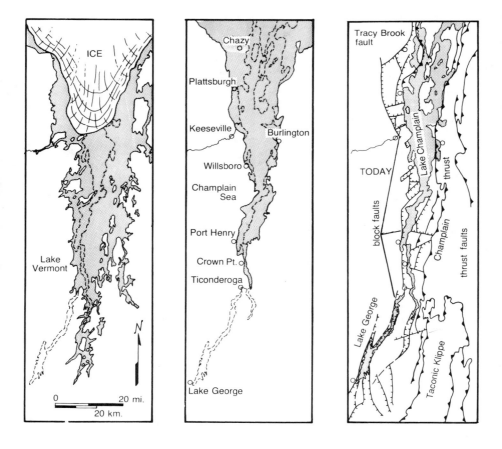

Three maps of lake stages in Champlain basin   Courtesy Kendall / Hunt Publishing Co.

poured in to form a saline embayment called the Champlain Sea, smaller than Lake Vermont but still larger than the present lake. Shoreline features, such as wave-cut terraces and beach strands, for each of these bodies of water remain visible inland from the present shore. They are now hundreds of feet higher than originally because the land surface has slowly rebounded after removal of the enormous weight of the continental ice sheet. Maximum rebound in New York state exceeds 500 feet along the St. Lawrence River. This same rebound is also partly responsible for the rocky banks of Lake Champlain, developed by wave erosion of the uplifted bedrock.

Blocks of Cumberland Head argillite "tile" with shale chips as "mortar," arranged by wave action

Stone-veneered Kensico dam, built in 1915

See maps pages 86 and 88.

Taconic State Parkway:
Kensico Dam—Interstate 90
100 mi./162 km.

The Taconic State Parkway is a beautifully landscaped "ribbon park" that runs south to north midway between the Taconic Mountains and the Hudson River. Between its southern terminus at the Kensico Dam and Peekskill (15 miles), it cuts diagonally across the north-northeast-trending metamorphic terrane of the Manhattan Prong. Between Peekskill and the Interstate 84 junction (20 miles), it crosses the ancient Hudson Highlands which, just across the Hudson River to the southwest, form the western boundary of the Triassic Newark basin. Between Interstate 84 and Interstate 90 (65 miles), it winds and weaves through the rolling, wooded hills and valleys of the frontal apron of the Taconic Klippe, or low Taconics.

The Kenisco Dam three miles north of White Plains is an unusual stone-veneered structure built in 1915 across the Bronx River as part of the New York City water supply system. The Kenisco Reservoir serves as a holding pond for water from the Ashokan Reservoir in the Catskills, 77 miles to the north-northwest. The Catskill aqueduct, which passes under the Hudson River, brings the water from Ashokan.

The course of the Bronx River, like most of the rivers in the Manhattan Prong, follows a narrow band of weak Inwood marble, named for good exposures in Inwood Park in northern Manhattan. The river follows the southwesterly trend of the marble and then turns southward to empty into the East River at the apex of Long Island Sound.

Between Kensico Dam and Hawthorne (3 miles), the road traverses hillier country of the Manhattan schist, then crosses Saw Mill River in another marble valley. Between Hawthorne and the New Croton Reservoir (10 miles), you first cross somewhat more rugged landscape developed on hard Fordham gneiss and then Manhattan schist. There are excellent large roadcuts near the contact between these two units four miles south of the handsome arch bridge over the reservoir, exposing banded and granitic gneisses with upended layering. Other large cuts in Manhattan schist are between there and the reservoir.

The boundary between the Manhattan Prong and Hudson Highlands crosses the Parkway from east to west, three miles north of the reservoir, about five miles east of Peekskill. Peekskill is on the north side of the Cortlandt igneous complex, an oval area roughly 10 miles from east to west and 5 miles from north to south. This is a series of funnel-shaped igneous bodies containing a wide range of dark rocks, some of which are quite rare. The complex was formed by multiple intrusions of molten rock, magma, during the late stages of the Taconian mountain-building event; each intrusion invaded the earlier ones so that locally the various rock types are intimately intermingled.

Commercial emery deposits exist near the contacts of the intrusions. Emery is a dense rock composed of a number of hard minerals such as magnetite, spinel, corundum, ilmenite, garnet, sillimanite, and cordierite. In addition to its use as an abrasive, it is embedded as a nonslip aggregate in stair treads and in floors of commercial buildings. This emery was formed from the Manhattan schist by "contact" metamorphism induced by the heat of the intruding magma and by chemical exchange with the magma. The emery, therefore, is chemically different from the unaltered Manhattan schist.

One mile north of NY 35/202, the Crompond road to Peekskill, you cross the Devonian Peekskill granite, which forms a small intrusive body, or "stock," in the Manhattan schist along its fault boundary with the Hudson Highlands on the northeastern side of the Cortlandt complex.

At midpoint between Peekskill and Interstate 84 the road crosses the deep little valley of Peekskill Hollow fault. The northern half of the Highlands traverse is wild, heavily wooded country dotted with numerous lakes and marshes and crossed by the Appalachian Trail.

Much of the region lies in the Clarence Fahnestock Memorial State Park. Just south of Interstate 84, the road descends into another deep fault-line valley across the face of Hosner Mountain with good views over the low Taconics.

Between Interstate 84 and the northern end of the Taconic Parkway at Interstate 90 (65 miles), the route cuts through a storm-tossed sea of rocks in the multiple landslides of the low Taconics. The Taconic mountains, or high Taconics, which rise to the east, are constructed not only of landslide debris, but also of highly contorted thrust fault "slices" that were shoved westward at the height of the Taconian mountain-building event. The topography here is a unique New York landscape. The numerous low hills have a distinctly hummocky appearance that reflects the crumpling and upending of their component, slabby, metamorphic rocks. The grade, or temperature, of metamorphism is generally low along the route but decreases slightly from south to north (see NY 22: White Plains—Interstate 90 discussion). Roadcuts near James Baird State Park 11 miles north of Interstate 84, for example, expose greenschists, colored green by the metamorphic mineral chlorite, and phyllite, with its distinctive pearly luster on finely wrinkled cleavage surfaces. The cuts farther north become progressively more slatey, displaying the telltale flat cleavage plates of this lower grade rock-type and black, gray, green, or red coloration. These metamorphic mineral zones trend north-northeasterly approximately parallel to the thrust faults and folds; the road cuts diagonally across the zones in a south-north direction.

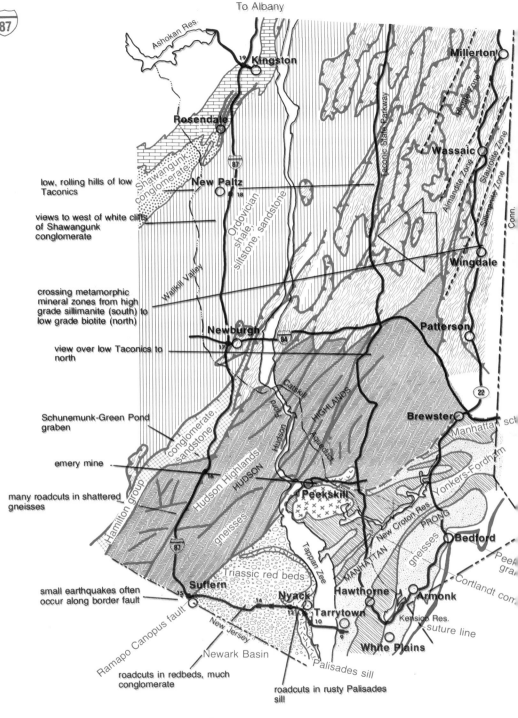

To Albany

87

Ashokan Res.

19 Kingston

Millerton

Rosendale

Biotite Zone

Wassaic

Almandite Zone

Staurolite Zone

Sillimanite Zone

Conn.

low, rolling hills of low Taconics

87

Shawangunk conglomerate

New Paltz

18

views to west of white cliffs of Shawangunk conglomerate

Ordovician shale, siltstone, sandstone

Wingdale

Wallkill Valley

Taconic State Parkway

crossing metamorphic mineral zones from high grade sillimanite (south) to low grade biotite (north)

Newburgh

84

17

Patterson

view over low Taconics to north

Catskill

22

Hudson

HIGHLANDS

Marbles

Brewster

Manhattan sch

Schunemunk-Green Pond graben

conglomerate sandstone

Hudson Highlands

HUDSON

Yonkers-Fordham

emery mine

Hamilton group

gneisses

Peekskill

New Croton Res.

PRONG

many roadcuts in shattered gneisses

87

gneisses

Bedford

Peek gra

Triassic red beds

Tappan Zee

MANHATTAN

Cortlandt con.

small earthquakes often occur along border fault

Ramapo Canopus fault

Suffern

15

14

Nyack

12

Tarrytown

10

Hawthorne

9

Armonk

suture line

Kensico Res.

White Plains

New Jersey

Newark Basin

Palisades sill

roadcuts in redbeds, much conglomerate

roadcuts in rusty Palisades sill

Interstate 87
TARRYTOWN—KINGSTON

Interstate 87:
Tarrytown—Albany
146 mi./236 km.

Tappan Zee is the name given to the extremely broad section of the Hudson River valley that stretches for 18 miles between Dobbs Ferry and Stony Point, a segment of the remarkable Hudson River estuary. High tides extend 160 miles upriver to Troy where the water surface is only about 10 feet above sea level. South of Dobbs Ferry, the river is more narrowly confined between the resistant metamorphic rocks of the Manhattan Prong on the east in Yonkers, the Bronx, and Manhattan Island and the Triassic Palisades sill on the west. North of Stony Point, the river carved a narrow, zig-zagging slot through the Precambrian Hudson Highlands to Cornwall-on-Hudson. Glacial erosion has made this into a true fjord, like those of Norway. A fjord is a steep-sided, coastal valley first carved by a stream, then by a glacier, then drowned by the sea. The broadness of the Tappan Zee reflects the weakness of the underlying soft shales, sandstones, and arkose of the Triassic Newark group, which were easily worn away by river and glacial erosion. The three-mile span of the Tappan Zee Bridge between Tarrytown and South Nyack is about the average width of this section of the river.

In the 13 miles between the Tappan Zee Bridge and Suffern, the highway traverses east-west across the nearly level Newark basin with its distinctive brick-red mudstones, shales, sandstones, arkoses, and conglomerates of Triassic age. The basin, only the northern tip of

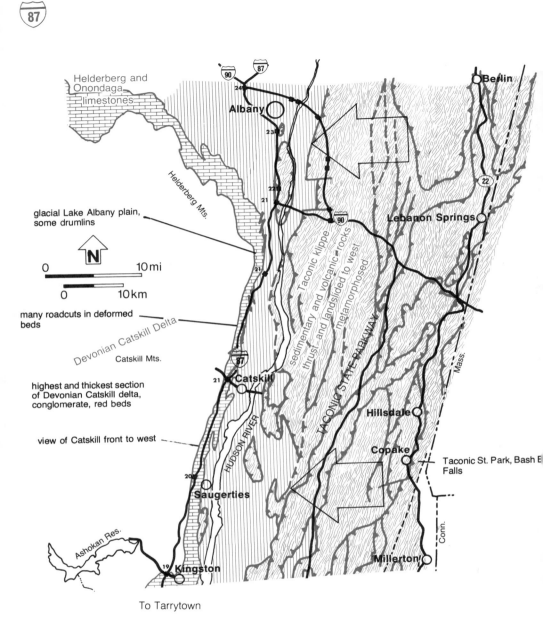

- Helderberg and Onondaga limestones
- glacial Lake Albany plain, some drumlins
- many roadcuts in deformed beds
- highest and thickest section of Devonian Catskill delta, conglomerate, red beds
- view of Catskill front to west

Devonian Catskill Delta

Catskill Mts.

Helderberg Mts.

HUDSON RIVER

Ashokan Res.

Taconic klippe

sedimentary and volcanic rocks thrust- and landslided to west metamorphosed

TACONIC STATE PARKWAY

Taconic St. Park, Bash E Falls

Mass.

Conn.

Albany

Berlin

Lebanon Springs

Hillsdale

Copake

Millerton

Catskill

Saugerties

Kingston

To Tarrytown

0   10 mi

0   10 km

N

Interstate 87
KINGSTON—ALBANY

which lies in New York, is one of several along the east coast of North America that dropped along block faults during Triassic-Jurassic time, about 200 million years ago and received sediments from the adjacent highlands. This is dinosaur country, but so far only footprints have been discovered in New York. The sediments thicken westward; at Suffern they achieve a maximum thickness of 20,000 feet at the Triassic border fault.

The high, flat-topped ridge on the east side of the basin by the river is part of the Palisades sill, a tabular mass of basaltic rock sandwiched within the sedimentary strata. Roadcuts reveal a rusty-weathered, highly fractured rock. The level landscape across the basin is a product of post-Triassic erosion of the redbeds, which dip, or slope, toward the border fault. In historical perspective, the "basement" rocks of the New York City group were by Triassic time reduced by erosion to a fairly level plain. The border fault, which is really an old fault, then gradually dropped the basin side 5000-8000 feet during Triassic-Jurassic time, and tilted the bedrock surface to the northwest, while the depression filled with sediments washed down from the upthrown block. Fault movement also tapped a deep source of magma which forced its way upward and then between the strata to form the Palisades sill and other similar bodies. In the Watchung Mountains of New Jersey, the same magma broke the surface to form lava flows.

Suffern is at the border fault, where the Hudson Highlands rise 800 feet above the plain in an impressive scarp. The dramatic change in landscape reflects the greater resistance, more complex geology, and longer history of the Precambrian gneisses, quartzites, and marbles. These are the deeply eroded remnants of a mountain range that rimmed the east coast of ancient North America from Labrador to Mexico during the Grenville mountain-building event, some 1100-1300 million years ago. The Adirondack rocks are part of this same mountain core. The modern Hudson Highlands in New York extend northeastward to the Connecticut border and southwestward into New Jersey. They are creased by a lot of linear, northeast-trending valleys eroded along fault-zones, many containing elongate lakes. For 12 miles north of Suffern, the highway passes through this rugged range along the Ramapo River, past large cuts in brownish, greyish, greenish, and buff-colored gneisses, some with well-defined layering, and most highly fractured.

Between mile 43 and Albany, a distance of about 100 miles, all of the visible bedrock is Paleozoic sedimentary strata with locally extensive folding and faulting. Between miles 43 and 47 are several low cuts in pale buff limestone or dolostone of the Wappinger group of Cambrian-Ordovician age. Between miles 47 and 55, you cross the

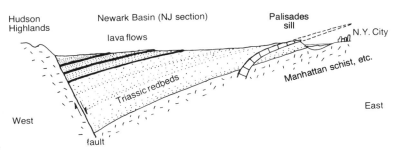

Cross-section, Newark basin with Palisades sill

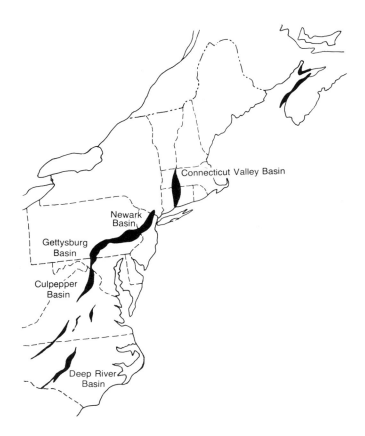

Triassic fault-block basins of the eastern United States

90

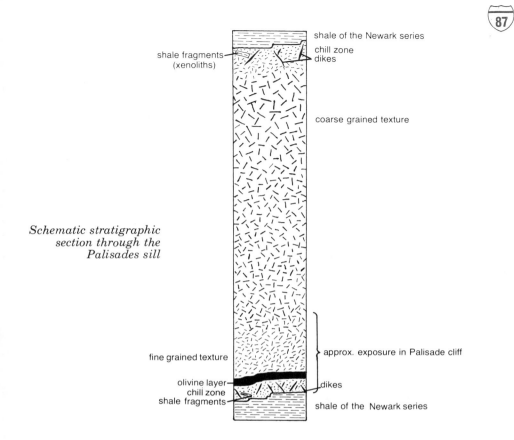

shale of the Newark series
chill zone
dikes
shale fragments (xenoliths)
coarse grained texture

Schematic stratigraphic section through the Palisades sill

approx. exposure in Palisade cliff
fine grained texture
olivine layer
chill zone
shale fragments
dikes
shale of the Newark series

Schunemunk-Green Pond graben floored by Silurian-Devonian conglomerate and sandstone. The time of graben faulting has not been established, but some geologists believe it was contemporaneous with the Newark basin.

Between mile 55 and Kingston, you follow the Wallkill Valley, a remarkably broad lowland about 65 miles long and 20 miles wide. Its western boundary is the Shawangunk Mountain scarp; the eastern boundary is Marlboro Mountain north of Newburgh, the Hudson Highlands on the south. The bedrock floor consists almost entirely of weak Ordovician shales and graywackes of the Normanskill and Martinsburg groups. Like so many other regions of New York State, the Wallkill Valley was extensively modified by Ice Age glaciation. It was a convenient channelway for ice lobes, and later, during glacial recession, it was a basin for a succession of proglacial lakes, including Lake Albany, and the sedimentation associated with them. The principal drainage now is the north-flowing Wallkill River, which heads in northern New Jersey and enters the Hudson near Kingston.

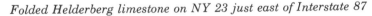
Near New Paltz, the east-facing Shawangunk scarp rises west of the highway and gradually converges with it from south to north. The ridge is held up by the Silurian Shawangunk conglomerate, a white, quartz-pebble conglomerate that stands up in vertical cliffs. The rock is well-indurated and bedded which, combined with the inward dip of the bedding on the cliff faces, makes the "Gunks" the best rock climbing region in the eastern United States. On almost any summer day, the cliff face for miles is practically crawling with climbers, whose cars are parked in every available space on nearby roads. The scarp dies out at Rosendale, north of New Paltz, at the limit of the conglomerate beds.

Between Rosendale and Albany, Interstate 87 follows the Taconic unconformity. Many roadcuts expose greyish Devonian Onondaga or Helderberg group limestones, while others display dark shales and siltstones of the overlying Hamilton group. All of the beds show some degree of deformation, dipping this way or that. A few cuts show anticlinal folds and/or faults. The deformation results from the uplift of the ancestral Acadian Mountains in the New England region during middle to late Devonian time.

Kingston is one of the gateways to the Catskill Mountains by way of NY 28. The Catskills are the namesake for the Devonian Catskill Delta where the sediments reach their greatest stratigraphic thickness. The imposing, high scarp of the mountains visible west of the highway, called the Catskill Mural Front, forms the eastern limit of the Allegheny Plateau. The "Wall," though heavily forested, has steps that reveal differential erosion of alternating hard and soft, nearly horizontal layers.

Catskill, at Exit 21, is another gateway to the Catskill Mountains, along NY 23 or 23A. Between there and Exit 21B, huge roadcuts in highly contorted Onondaga/Helderberg limestones show upfolded

Folded Helderberg limestone on NY 23 just east of Interstate 87

anticlines and downfolded synclines and, in some places, shales with nearly upright beds. The entire route between Kingston and Catskill lies close to the western limit of the Taconic Klippe (see NY 22 discussion).

The broad lowland east of the highway between Exits 21 and 21B is a terrace of the preglacial Hudson River into which the modern river has cut its valley. It is also part of the Lake Albany plain, and several drumlins project above its flat surface.

Between Exits 21B and 24 to the Adirondack Northway, the route traverses the Lake Albany plain, and passes over erosional outliers of the Taconic Klippe, containing Austin Glen graywackes and shales. The Devonian contact, meanwhile, veers away from the highway to the northwest, curving around the Helderberg Mountains. Note the large limestone quarry at mile 132. This region is also dotted with quarries in the "varves," or glacial lake clays that are used in brick-making and ceramics, along with ubiquitous quarries in sand and gravel.

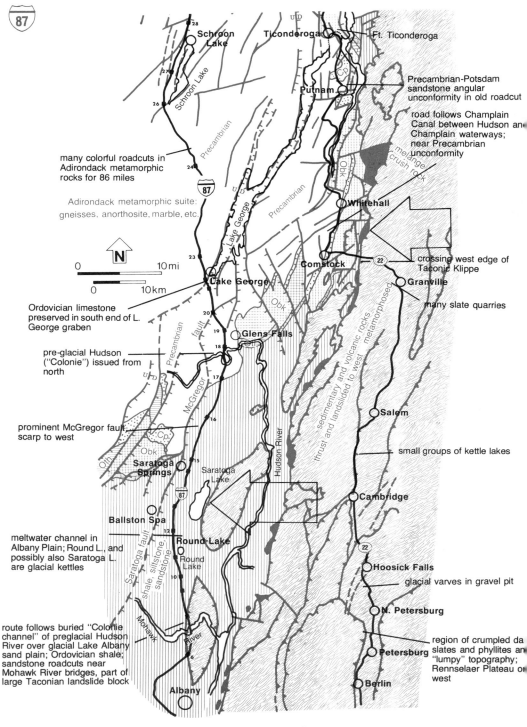

Interstate 87
ALBANY—CANADA BORDER

many colorful roadcuts in Adirondack metamorphic rocks for 86 miles

Adirondack metamorphic suite: gneisses, anorthosite, marble, etc.

Precambrian-Potsdam sandstone angular unconformity in old roadcut

road follows Champlain Canal between Hudson and Champlain waterways; near Precambrian unconformity

crossing west edge of Taconic Klippe

Ordovician limestone preserved in south end of L. George graben

many slate quarries

pre-glacial Hudson ("Colonie") issued from north

prominent McGregor fault scarp to west

small groups of kettle lakes

meltwater channel in Albany Plain; Round L. and possibly also Saratoga L. are glacial kettles

glacial varves in gravel pit

route follows buried "Colonie channel" of preglacial Hudson River over glacial Lake Albany sand plain; Ordovician shale; sandstone roadcuts near Mohawk River bridges, part of large Taconian landslide block

region of crumpled dark slates and phyllites and "lumpy" topography; Rennselaer Plateau on west

Schroon Lake, Ticonderoga, Ft. Ticonderoga, Putnam, Whitehall, Comstock, Granville, Lake George, Glens Falls, Salem, Saratoga Springs, Saratoga Lake, Cambridge, Ballston Spa, Round Lake, Hoosick Falls, N. Petersburg, Petersburg, Berlin, Albany

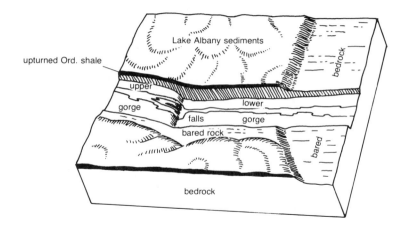

Schematic block diagram of Cohoes Falls. Upper gorge was carved earlier when proglacial Lake Fort Ann filled the Hudson Valley (right side) and provided a base level of falls erosion. New falls formed later when the lake drained and have since receded 2000 feet. Adapted from Stoller (1918).

Interstate 87:
Albany—Canada Border
176 mi./285 km.

Between Albany and Glens Falls, a distance of 50 miles, Interstate 87 traverses the glacial Lake Albany sand plain. Some of the sand has been reworked by wind to form dunes, now arrested in their migration by vegetation or human development. Parts of the plain are creased with meltwater channels excavated during the waning stages of the Ice Age. Between Exits 6 and 8 near Cohoes, the road crosses the Mohawk River over one of two attractive twin arch bridges. On the approach to these are long cuts in dark Ordovician graywackes and shales. Downstream, at Cohoes, the river makes a final plunge of about 65 feet over Cohoes Falls before joining the Hudson at Albany. During most of the year, the river is not much more than a trickle, because the water is diverted for power generation; but in spring and fall, it often becomes a roaring cataract, truly one of the natural spectacles of upstate New York. The lower gorge was carved by recession of the present falls in post-Lake Iroquois

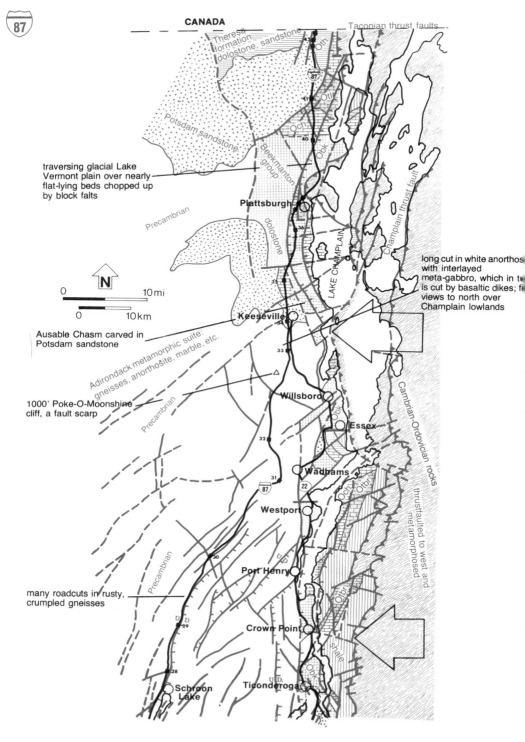

CANADA

Taconian thrust faults

Theresa formation dolostone, sandstone

Potsdam sandstone

traversing glacial Lake Vermont plain over nearly flat-lying beds chopped up by block falts

Beekmantion group

Plattsburgh

dolostone

LAKE CHAMPLAIN

Champlain thrust fault

long cut in white anorthos with interlayed meta-gabbro, which in tu is cut by basaltic dikes; fi views to north over Champlain lowlands

N

0   10 mi
0   10 km

Precambrian

Keeseville

Ausable Chasm carved in Potsdam sandstone

Adirondack metamorphic suite: gneisses, anorthosite, marble, etc.

Precambrian

Willsboro

Essex

Cambrian-Ordovician rocks

1000' Poke-O-Moonshine cliff, a fault scarp

Wadhams

87   22

Westport

Precambrian

thrustfaulted to west and metamorphosed

many roadcuts in rusty, crumpled gneisses

Port Henry

shale

Crown Point

Schroon Lake

Ticonderoga

Interstate 87
ALBANY—CANADA BORDER

time. The falls have thus far receded 2000 feet from the shoreline of glacial Lake Fort Ann, a successor to Lake Albany, but at a lower water level.

Black crumpled rocks exposed near the twin arch bridges and in the Cohoes Gorge are part of an erosionally isolated mass that moved in the gigantic landsliding that accompanied the Taconian mountain-building event in Ordovician time, 445-435 million years ago. The original sea muds were deformed while still soft, and have since been transformed to solid rock. Erosion of the Cohoes Gorge has planed off the upturned edges of these beds.

This Albany region is the site of major drainage changes during Pleistocene glaciation. The twin arch bridges, for example, are adjacent to the deeply buried, north-south "Colonie channel" of a preglacial river that does not exist today. The bottom of the bedrock gorge now lies some 200 feet below the modern Mohawk River. At the time of the Colonie Channel, the ancestral Mohawk lay several miles south of its present position, and joined the Hudson River south of Albany, at about 100 feet below the present sea level. The Hudson during that time followed a course north of Albany similar to the present one, but deeper.

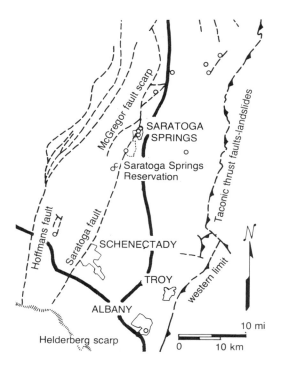

Up-down faults in the Saratoga Springs-Albany region showing location of "springs" (actually wells–open circles), and the western limit of Taconic thrust faults–landslides

Lake George from Prospect Mountain, with flanking hills

A bedrock ridge just east of the highway on the north side of the Mohawk River protrudes about 100 feet above the sand plain. This ridge was an island in Lake Albany, which here had a shoreline at about 300 feet above sea level, about the level of the sand plain.

Just north of Exit 11 to Round Lake, the road crosses one of the glacial meltwater channels now occupied by Ballston Creek. Nearby to the south, Round Lake and Little Round Lake, as well as Saratoga Lake farther north, are possible kettle lakes lying in glacial debris which now fills the preglacial Colonie channel. The highway follows this channel almost to Glens Falls.

Between Exits 12 and 13, the road crosses the valley of Drummond Creek past large roadcuts in black Snake Hill shale that underlies the sand plain. Between Exits 17 and 18 (Glens Falls), you cross the Hudson River.

Between Saratoga Springs and Lake George, a conspicuous scarp lies west of the highway. This is the fault-line scarp produced by differential erosion of the McGregor fault, which forms the western boundary of the Lake George depression. faulting here corresponds in age and origin to that of the Little Falls and Noses faults discussed in Interstate 90: Syracuse-Massachusetts border. These are up-down

98

of eroded Precambrian rocks. Richard K. Dean photo, courtesy Kendall / Hunt Publishing Co.

"block" faults that chop up the southern and eastern sides of the Adirondacks and adjacent regions. Downthrown blocks or "grabens" are flanked by parallel upthrown blocks. The Lake George graben is a good example. Others hold Sacandaga Lake and Lake Champlain.

The McGregor fault controls the locations of many of the carbonated mineral springs of the Saratoga region, the reason for its development as a health spa. The vertical displacement along the McGregor fault is estimated at 2800 feet at Lake George. The upthrown block you see along this route is Precambrian rock, 1.1 billion years old. In Battleground State Park at the south end of Lake George and on some of the islands in the lake, there are exposures of Ordovician limestones on the floor of the graben. Ordovician rocks probably covered most of the Adirondack region before block faulting and the much later doming of the modern mountains. Erosion has removed these rocks from the high places, but not from some depressions like the graben floors.

Lake George is uniquely beautiful, a long, slender lake nestled between heavily forested high mountains, the upthrown blocks, or "horsts," that border the graben. The crystal clear waters for which it is best known come from short feeder streams that drain hard bedrock, and thus carry little sediment. The lake is studded with 184 lovely wooded islands, mostly in the shallow central part called the "Narrows."

In preglacial, Tertiary time, the lake basin was the site of two river valleys, one flowing northward from a drainage divide at the Narrows, the other southward. Glaciation broadened and deepened the valley and reduced the Narrows divide to a low level. During Wisconsin glacial recession, the southern outlet, east of French Mountain across the bay from Lake George village, was blocked by moraine and drumlins, setting the stage for filling of the basin. The lake now drains into Lake Champlain via tiny Ticonderoga Creek at its northern end.

Between Exits 20 to Lake George and 34 to Keeseville, a distance of 86 miles, the highway passes through the eastern edge of the Adirondacks from the McGregor fault scarp to the boundary with the Champlain Lowlands. Cuts in Precambrian gneisses and anorthosite are numerous and large. The blend of these rugged, and often colorful, rock cuts, with grass slopes and natural tree cover, combined with the superb mountain scenery led to the designation of the Northway as one of America's most scenic highways. It would be futile to try to describe the geology of every one of these cuts; stopping to take a close look at them is illegal anyway. In general, however, the road between Exit 21 at Lake George and Exit 28 north of Schroon Lake (36 miles) goes through a smorgasbord of complexly folded, grayish, pinkish, or brownish gneisses and ocasional black gabbroic rock. In several places, the rocks are cut by block faults, and, as a result, are highly fractured and rusty from oxidation of the iron-bearing minerals in the crush zone.

Between Exits 26 and 30 (60 miles), the route follows the floor of a narrow graben that transects the eastern edge of the Adirondack anorthosite body (see Whiteface Mountain Memorial Highway discussion). The anorthosite typically appears white or greenish white. Locally the extensive fracturing reveals the proximity of the faults that border the graben. One cut between Exit 29 and 30 is brick red, again as a result of crushing and oxidation of a more iron-rich rock.

Between Exits 30 and 31, the road again passes through a variety of gneisses including gabbroic types; and some are rusty. Anorthosite, however, holds sway over most of the way between Exits 31 and 34. Between Exits 32 and 33, you will pass close to Poke-O-Moonshine Mountain with its spectacular 1000-foot high cliff, a favorite among the really serious Adirondack rock climbers. The cliff is on the upthrown side of the north-south-trending Poke-O-Moonshine fault and consists of highly resistant granitic gneiss, whereas the downthrown

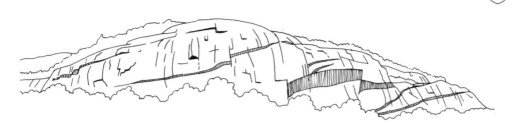

Thousand-foot high Poke-O-Moonshine cliff, an upthrown fault of granitic gneiss with dark sills of basaltic composition offset by small faults
Courtesy Kendall/ Hunt Publishing Co.

side is made up of rocks belonging in the anorthosite suite. Several nearly level bands of dark, highly fractured gabbroic or basaltic rocks crease the face and are offset in a few places by vertical faults. Most of the established climbing routes, in part, follow ledges at the bases of these bands. Otherwise the cliff face has been scoured quite smooth by glaciation. The steepness and suddenness of the precipice is best appreciated in the approach from Exit 33 on US 9.

An interesting long cut occurs just north of Exit 33 on the west side of the north-bound lane. The cut exposes whitish metanorthosite with a broad, gently sloping, band of black metagabbroic rock. The band is itself transected by several thin, non-metamorphosed, basaltic dikes. The geologic history revealed here is that of a deeply buried anorthosite being intruded along a layer boundary or fracture by gabbro, then later being subjected to metamorphism. This was followed by a long period of erosion which brought the rocks close to the surface. Finally, the basaltic dikes were intruded along new fractures, in essence repeating the much earlier history of igneous activity.

At the top of a rise about one mile north of Exit 33, there is a splendid panorama of the Champlain Lowlands, which begin about Exit 34 to Keeseville. The transition from the rugged Precambrian terrane into that of the stratified sedimentary rocks of Cambrian and Ordovician age is dramatic. Keeseville lies just south of the famous Ausable Chasm, a narrow, postglacial gorge of the Ausable River cut into a thick section of Potsdam sandstone (see NY 9N: Ausable Chasm—Keene discussion). For the most part, the rock strata from here all the way to the Canadian border near Champlain are flat-lying or nearly so, but chopped up by numerous block faults that border the Lake Champlain depression.

The rest of the way to the border there are very few roadcuts and the landscape is quite flat and open. The bedrock strata actually dip this way and that as a result of the block (and locally thrust) faulting;

101

Cumberland Head (Trenton group) argillite by ferry dock; this unit has been thrust-faulted (pushed) westward.

but extensive glacial scour, coupled with the later sedimentation in Lake Vermont and the Champlain Sea, have leveled the surface. Note the abundance of sandy soil along the way. Several streams that cross the plain, like the Ausable at Ausable Chasm, the Salmon south of Exit 6 to the Plattsburgh Air Force Base, and the Saranac at Plattsburgh, have cut down into it to reach the Lake Champlain baselevel, producing bedrock gorges.

See maps pages 86 and 88.

NY 22:
White Plains—Berlin
151 mi./245 km.

Good exposures of pinkish Yonkers gneiss and darker, crumpled Fordham gneiss are visible in roadcuts near the Kensico Dam over the Bronx River 3 miles north of White Plains. The dam, itself, is an architectural wonder with an unusual veneer of massive stonework. Built in1915 as part of the New York City water supply system, the Kensico Reservoir behind it is used as a holding pond for water from the Ashokan Reservoir of the Catskill Mountains. The connecting link is the 77-mile long Catskill aqueduct that passes under the Hudson River near West Point.

Between Kensico Dam and the Interstate 84 junction (35 miles), the Manhattan schist is exposed in several roadcuts. Over most of the way, the route also lies just a few miles west of a profound geologic boundary, a suture line begun during the Taconian mountain-building event when offshore volcanic islands were shoved up against the continent. Final welding occurred later during the Acadian uplift as ancestral Africa collided with North America. The boundary is marked by numerous bodies of serpentinite, a dense, greenish, greasy-looking rock composed of the mineral serpentine, and formed by alteration of dark oceanic crustal materials that were squeezed up along the juncture. Best exposures of the suture line are on Staten Island, Manhattan Island, New Rochelle, and Port Chester.

Between the Interstate 84 junction and Patterson (10 miles), you cross the Hudson Highlands, with roadcuts in intensely deformed banded gneisses. The Highlands are part of the Precambrian basement rock of New York, which also includes the Fordham gneiss and the Adirondacks.

Between Patterson and Interstate 90 (82 miles), geology and topography are dramatically different from that to the south. The hills and valleys trend north-south and reflect the underlying geology faithfully. The mountains are not very high, but are generally steep and heavily wooded.

The route follows a system of valleys developed on carbonate rocks, whereas the adjacent hills, for the most part, are held up by crumpled schists derived by the metamorphism of original shales and graywackes. Between Patterson and Millerton, for example, you will see several roadcuts in the greenish-gray Cambrian-Ordovician Stockbridge marble, and a few exposures of dark schists.

This southern end of the Taconics also displays the steepest metamorphic gradient of the entire range. There are low temperature phyllites and slates on the west, which grade rapidly eastward, over about 10 miles, into high temperature schists. Metamorphic grades are recognized by the appearance of certain index minerals that crystallize during the metamorphism. Green, soft, micaceous chlorite, for example, is distinctive for low grade rocks, but will only appear if the bulk chemistry of the affected rock is suitable. The fine-grained white mica that provides the pearly sheen to the cleavate surfaces of phyllite represents a slightly higher grade. The appearance of black biotite mica signals the next higher grade, which is normally accompanied by a coarser texture as a result of the more intense metamorphism. The presence of the almandite garnet in the micaceous schists, signals a higher grade yet. Then comes staurolite, and sometimes kyanite, usually in large elongate crystals set in a finer, biotite-rich matrix. Finally, the sillimanite, representing the highest grade in this sequence, shows up as tiny needles that sparkle in the sunlight. Such a gradation from low to high temperature metamorphic minerals is referred to as a progressive metamorphic sequence.

Between Wingdale and Wassaic (12 miles), you cross from sillimanite-grade rocks to staurolite-grade rocks. The valley here is particularly broad, and East Mountain looms large on the east side of it. The mountain, which rises about 700 feet above the valley, is held up by hard gneisses and quartzite. Just south of Wassaic, you pass through the narrow water gap of Wassaic Creek, cut in a high ridge supported by resistant schists. In so doing, you also cross a major thrust fault which shoves the staurolite-bearing rocks of the east side over the almandite-grade rocks of the west side.

Between Wassaic and Millerton (13 miles), the road again follows a lowland carved in the Stockbridge marble. The almandite-biotite boundary is crossed just north of Millerton. Therefore, the route has passed over four metamorphic mineral zones in the short distance of 25 miles. Since the zones trend north-northeast parallel to the structural grain of the mountains, the east-west distance across them is much less, only about 5 miles. This is a remarkably rapid change that appears to result in large part from the stacking of crustal plates by the westward-directed thrust faults. Metamorphic mineral zones that once occupied a much broader region have thus been shoved together into a narrow band. The metamorphic gradation becomes much less pronounced to the north around the slate belt of New York and Vermont.

Between Millerton and Hillsdale (19 miles), the road follows another broad, open, carbonate valley, principally in the Ordovician Beekmantown group dolostone and the Copake formation, but with a few roadcuts in phyllites. Many such valleys have flat floors because they served for a time as meltwater lake basins during Wisconsin deglaciation, and were partly filled with sediments from inflowing streams. It is common in the Taconic valleys for bedrock knobs to project through the sediment fill, producing a hummocky topography.

Bash Bish falls in Taconic State Park

Phyllite roadcut near New Lebanon

Taconic State Park lies in the high hills on the east side of the valley between Millerton and Hillsdale. This impressive mountain rampart, which rises more than 1600 feet from the valley, is sustained by metamorphosed shales and graywackes in an erosionally separated thrust slice, or klippe, whose western base lies near the foot of the slope visible from the road. Entrance to the park is at Copake Falls, 4 miles south of Hillsdale, where Bash Bish Creek has carved a deep slot in the mountain front.

Between Hillsdale and the Interstate 90 junction (19 miles), the route weaves its way through a maze of small klippen.

Between Interstate 90 and Berlin (25 miles), the road follows a "window" first exposing gray, thick-bedded, and moderately-deformed marble of the Stockbridge formation in a few roadcuts. The flanking hills are each capped by erosional remnants of a single fault slice of phyllite and minor quartzite. An excellent place to see these rocks and to overlook the Taconic landscape is on the Lebanon Road, US 20, which goes over the mountain east of New Lebanon. The northern segment of the window between New Lebanon and Berlin is underlain by Walloomsac formation schists.

106

See map page 94.

NY 22:
Berlin—Comstock
68 mi./110 km.

This section of NY 22, with the exception of about 4 miles near Comstock, lies entirely within the Taconic Mountains sub-province of the New England Uplands, geologically known as the Taconic Klippe. The region is characterized by a complex array of north-northeast-trending thrust fault and gravity slide slices that were bodily transported westward during the Taconian mountain-building event and stacked against each other like so many folded bridge chairs leaning haphazardly against a wall. Subsequent deep erosion has formed hills from many of the slices of more resistant materials. Valleys have largely been excavated in crush zones along faults or in weaker marbles, and some have actually worn completely through the klippe so as to form "windows" that expose the underlying, in-place bedrock, or lower slices. It is obvious that the landscape is strongly controlled by geology. In general, the Taconic hills are steep, uniformly wooded, and locally have a rather "lumpy" surface texture that reflects intense folding, faulting, and cleavage of their thinly laminated bedrock. Unlike NY 22 south of Interstate 90, this northern traverse of the Taconics almost exclusively encounters low grade metamorphic rocks, such as slates, phyllites, and metagray-wackes. At the northern end, the route passes through the famous Granville slate belt.

Much of the difficulty geologists have in working out the complex geology of the Taconic Klippe has been in trying to distingush erosional and fault/gravity remnants of rock units of certain age and lithology from place to place. In the southern Taconics, this is immensely complicated by the rapid increase in metamorphic grade from west to east (see NY 22: White Plains—Interstate 90 discussion). Where the grade is low, fossils may be preserved, and separate outcrops can sometimes be correlated by similar fossil content. Higher grade metamorphism almost invariably destroys fossils.

Between Berlin and North Petersburg (11 miles), the highway is narrowly confined between the Rennselaer Plateau to the west, with graywacke caprock, and Berlin Mountain of the Taconic Range proper to the east, with phyllite and minor quartzite caprock. The road here follows the Little Hoosic River, which joins the Hoosic at North Petersburg.

Between North Petersburg and Eagle Bridge (12 miles) is the broad, flat valley of the Hoosic River, filled with both stream and glacial lake sediments. A gravel pit by the west side of the road near Hoosic reveals tell-tale glacial varves, those peculiar, laminated, grayish, clay deposits of glacial lakes. A distinctive feature of varves is their annual pairing—thin, blackish organic-rich layers formed in winter months coupled with thicker, lighter silty layers deposited in the summer months. Thus, the years of accumulation can be counted like tree growth rings. Glacial meltwater is almost always milky, laden with rock flour generated by the grinding together of rocks in and adjacent to the ice. The flour is kept in suspension by the turbulence of the streams but slowly settles out in the still water of lakes. Warm period deposits are thicker because more flour is released as the ice melts.

Between Eagle Bridge and Salem (17½ miles), the road passes diagonally across a stacked "card deck" of fault slices with infrequent

Glacial varves near Hoosic.

roadcuts in dark, deformed slates. This is a broad valley with lovely, rolling farmland and open views of the mountains to the east. A small group of lakes halfway between Cambridge and Salem, including Lake Lauderdale, Schoolhouse Lake, Dead Pond, and Hedges Lake, are probably kettles. A few, poorly formed drumlins top the flat surface south of Salem.

Between Salem and Granville (17 miles), the road follows shallow valleys and climbs over low, hilly country with frequent open views. There are almost no roadcuts, but the region is well-known as the Vermont/New York, or Granville, slate belt. Numerous hills are dotted with small slate quarries, most of them now inactive. Slate production in New York is an old-time industry that dates back to the 19th century. It is now in its declining years, owing to antiquated, costly production and milling practices and the competition from synthetic substitutes.

Slate is formed by the low-grade metamorphism of shales. It is an extremely durable rock with the unique property of very flat, closely-spaced cleavage that makes it ideal for roofing and floor tiles. Another valuable asset is that it comes in a wide range of attractive colors—black, reds, greens, browns, grays—which are sometimes mixed on cleavage surfaces. By an accident of nature, all of the red slate of this region is found in New York. Thick slabs and blocks are also produced for use as flagging and dimension stone; crushed slate is sold for roofing granules.

The slate is quarried into thick slabs by drilling 1-inch holes along the cleavage and blasting with black powder. The slabs are then loosened and pried apart with crowbars and sledge hammers. Select pieces, averaging only 30% of the total, are transported to the mill. The rest is waste. At the mill, the slate may be further split with a jackhammer and sawn into desired dimensions. For making floor tile, slabs are sized to specific square and rectangular shapes. "Splitters" in the mill then cleave these slabs with a wide splitting chisel and hammer into sheets slightly thicker than ¼ inch. About half of the sheets are discarded for irregularities. The final step is gauging, whereby the split sheets are placed on a conveyor belt in a horizontal milling machine that shaves one face down, giving the tile a uniform ¼-inch thickness. The resulting flat milled surface also permits tile to be laid with an all-purpose flooring cement, as in the case of vinyl floorings. Total wastage from quarry to final product is about 85%; in Vermont, it averages 95%. Obviously, survival of the slate industry hinges on reducing this wastage. The tile is commonly marketed in cartons containing mixed colors and sizes, which can be set in various mosaic patterns.

Between Granville and the junction with US 4 near Comstock (12 miles), the route cuts east-west across the western boundary of the Taconic Klippe and enters the block-faulted Cambrian-Ordovician terrane of the east side of the Adirondacks. A few roadcuts near Comstock display gray Beekmantown dolostone of the Adirondacks. At Comstock, you cross the Champlain Canal which connects the southern end of Lake Champlain with the Hudson River at Fort Edward. At the US 4 junction, the road lies on a narrow strip of Potsdam sandstone next to the Grenville-age metamorphic rocks of the Adirondacks.

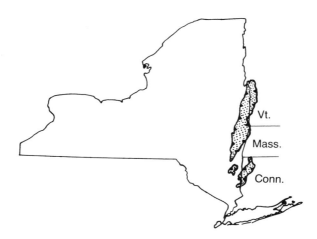

Taconic klippen (three shown) of eastern New York and adjacent New England states

See maps pages 94 and 96.

NY 22:
Comstock—Keeseville
90 mi./146 km.

THE CHAMPLAIN REGION

This route follows close to the shore of Lake Champlain along the eastern side of the Adirondack Mountains. It is a region of numerous block faults formed during the Taconian mountain-building event, dislocating both the Cambrian-Ordovician bedrock of the Champlain basin and the Precambrian rocks of the mountains.

Although this section is north of the Taconic Klippe, the Vermont side of the lake is cut by several major thrust faults that were pushed westward during the Taconian mountain-building event. The difference is that these thrust fault slices have not been isolated by erosion. The Champlain thrust fault, follows the full length of the eastern shore of the lake, and several other major thrust faults cut through the Champlain Islands. At least one projects into New York at the Cumberland Head peninsula near Plattsburgh. Along the western margin of thrusting, block faulting is prevalent. The block faults appear to result from the tensional stresses affecting the east coast of North America in late, or post-Taconian time during crustal stretching that preceded the Acadian uplift.

Between Comstock and Ticonderoga (33 miles), the Precambrian and early Paleozoic bedrock is diced by block faults, some trending generally north-south and others east-west. The cross-cutting faults

111

are probably related the the Champlain thrust. This is a common feature caused by differential resistance to thrusting from place to place. Where movement is severely obstructed, a slice of the thrust sheet may be held back, while the unimpeded segments on either side move ahead along wrench faults.

Between Comstock and Whitehall (7 miles), the route follows the Champlain canal in a rather broad and flat valley. This is part of the sediment plain of Lake Vermont, the proglacial lake that filled the Champlain basin as the Champlain lobe of the Wisconsin ice sheet retreated northward, but the glacier still blocked the St. Lawrence Valley. The southern tip of the lake was near Comstock; to the north, its waters extended far east and west beyond the present shoreline of Lake Champlain. See Interstate 87: Albany—Canada Border discussion.

The route here also lies close to the boundary between the Precambrian Adirondack gneisses, exposed in some large roadcuts, and the basal unit of the Paleozoic sedimentary strata, the Cambrian-Ordovician Potsdam sandstone.

West of Whitehall, the highway crosses South Bay in a down-dropped fault block, or graben, that extends northward at least 10 miles. This is just one of several graben structures in the region, including the much larger Lake George graben about 5 miles to the west.

Between South Bay and Ticonderoga (21 miles), the road continues along the narrow, 3- to 5-mile wide divide between the Lake George and Champlain basins. Numerous roadcuts display a variety of dark, light, and banded Precambrian gneisses, with cross-cutting, very coarse-grained, pinkish, pegmatite dikes. Most of the cuts are extensively fractured by the many faults mentioned earlier. There are also a few exposures of flat-lying Potsdam sandstone beds. The landscape reflects this complex geology and the road goes up and down over upthrown or downthrown fault blocks. A rest stop near Putnam is by huge cuts in gneiss with upended banding, and a splendid overview of the southern, river-like, end of Lake Champlain. Nearby on the Putnam entry road, there is a tree-shrouded outcrop in which this same banded gneiss is truncated by level Potsdam sandstone beds. The gneiss at one end has been sapped by weathering and erosion, leaving the sandstone overhanging. By a trick of fate, a large remnant of the gneiss remains suspended from this "ceiling," as if it were stuck there with glue. This is an unconformity almost identical to that of the Alex Bay cut on the northwest side of the Adirondacks, described in NY 37, 12, 12E: Ogdensburg—Cape Vincent—Watertown. It represents a gap in the rock record of about 600 million years.

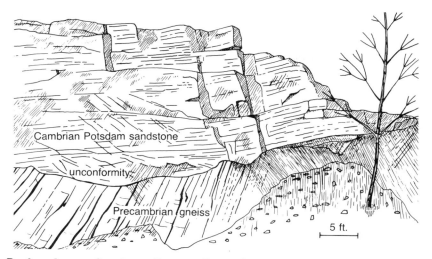

Profound unconformity at Putnam Center between uptilted and eroded Precambrian gneisses and flat-lying Potsdam sandstone beds. A nearly 600 million year time gap is represented. The exposure is similar to that of the "Alex Bay cut" on the other side of the Adirondacks (see NY 37: Ogdensburg—Cape Vincent—Watertown).
Courtesy Kendall / Hunt Publishing Co.

Near Ticonderoga, the road skirts Mt. Defiance, hard by Lake Champlain, with views of Fort Ticonderoga north across the bay. The fort, now completely restored, stands on a rocky rib that juts into the lake, creating a U-shaped bend and a narrow pass. Along with Fort St. Frederick at Crown Point, it was an important stronghold in the Colonial and Revolutionary wars, especially against waterborne invasion from Canada. The fort was held by the British at the outbreak of hostilities in 1775. Shortly thereafter, it was easily captured in a pre-dawn, bloodless raid by a band of about 83 "Green Mountain Boys" from Vermont under the joint command of Ethan Allen and Benedict Arnold. The Americans held it until 1777, when it was retaken by British General Burgoyne in a most unusual way. Burgoyne had his men haul cannon laboriously up the steep, wooded slopes of Mt. Defiance, where they were trained on the fort. Seeing their position defenseless, the Americans, under General Arthur St. Clair, stole away in the night. Burgoyne took possession on July 4, ironically exactly one year after the adoption of the Declaration of Independence! Burgoyne continued his push southward until he was defeated at the Battle of Saratoga in October in one of the most decisive battles of the war.

The Ticonderoga topograpy is, of course, geologically controlled. The bedrock of the fort promontory, and across Ticonderoga Creek to the foot of Mt. Defiance, is Ordovician limestone and dolostone of the Chazy, Beekmantown, and Black River-Trenton groups. The backbone of Mt. Defiance is Precambrian gneiss standing about 700 feet higher. Obviously, there is a fault across the northern end of the mountain along which the gneiss is upthrown.

Exposures of Paleozoic bedrock, including the basal reddish Potsdam sandstone, may be seen along Ticonderoga Creek near the village. This rather insignificant stream that drops about 225 feet to Lake Champlain is the only outlet for 33-mile long Lake George. From here, Lake George outflow continues northward to the St. Lawrence, then on to the Atlantic. Thus, waters that begin at the south end of the lake, a stone's throw from the Hudson River, must take a circuitous 650-mile route to the ocean, or the Gulf of St. Lawrence. The shorter Hudson route is blocked by the Glens Falls moraine, which is also the natural dam that made possible the filling of the lake basin.

Between Ticonderoga and Crown Point (8 miles), you cross the Lake Vermont sediment plain, with sandy deposits evident in several quarry pits. The sharp break with the mountain slope to the west marks yet another block fault where the Precambrian rocks are upthrown, this one extending north from Ticonderoga for a full 35 miles nearly to Willsboro. The mountainside is locally terraced by wave-cut benches and beach strands marking the high stands of Lake Vermont and the Champlain Sea. Glacial rebound has placed these shorelines well above their original levels. A good measure of at least one level of Lake Vermont is a huge outwash delta about halfway between Ticonderoga and Crown Point, that has sustained a large commercial sand and gravel operation. The flat upper surface, which is pitted with kettles, is about 525 feet above sea level, 385 feet above the road, and 430 feet above Lake Champlain.

This is a mining country, where a variety of ores were taken from the Precambrian hills in the past. There are, for example, numerous old graphite mines around Hague and Graphite west of Lake George, and Chilson Hill near Ticonderoga. Magnetite iron ore has been taken from the areas around Mineville, Clintonville, Harkness, and Port Henry. Presently, the only active mine is that of the wollastonite deposits near Willsboro. Wollastonite is a white calcium silicate mineral used principally in ceramic wall tile and porcelain, and as a paint extender. Garnet, used as an abrasive, is recovered as a by-product of the wollastonite operation (see NY 28: Warrensburg—Blue Mountain Lake discussion).

Between Crown Point and Willsboro (35 miles), the road continues close to the faulted Precambrian-Paleozoic contact. By Port Henry, the lake shore follows the fault and the road goes up and down over a

Canojarie shale

Chazy limestone

Up-down fault at Essex, with Chazy limestone (right) and drag-folded Canajoharie shale (left)

hilly landscape of gneisses and marbles. The ruggedness of the Precambrian landscape reflects the complex deformation and variety of rock types, and the more vigorous erosion of the upthrown sides of the faults. The downthrown blocks are floored by regularly stratified rocks that have been gently tilted this way and that by the block faulting. These have been glacially planed off and blanketed with flat-lying Lake Vermont and Champlain Sea sediments. The surface is punctuated here and there with low drumlins.

Willsboro, incidentally, is the site of British General Burgoyne's encampment on June 20-25, 1777, prior to his brilliant capture of Ticonderoga.

Block faults are exposed at several places along the rocky shore near Essex, south of Willsboro, but they are best seen from a boat. One just south of the Essex ferry dock places Chazy group limestone against blackish, shingled, Canajoharie shale, which represents a higher, and younger, stratigraphic level. A little farther south, Split Rock Point is an upfaulted block of Adirondack anorthosite and metagabbro that juts into the lake; the near shore of Whallon Bay is underlain by Trenton-Black River limestone on the downthrown, north side.

East of Willsboro on Jones Point is an unusually flat, glacially polished Potsdam sandstone surface that is now part of a private estate. The surface is so even, in fact, that it is used in its natural state as a tennis court in one place, and roadbed elsewhere. Glacially polished and striated surfaces are best preserved in the Potsdam sandstone because it is so hard and resistant to weathering. Scour surfaces on other sedimentay rock types and feldspar-bearing gneisses, dark amphibolites, and marbles of the Adirondacks, quickly deteriorate when exposed to the air.

Willsboro Point, north of the village, is a long, north-south peninsula made up of Beekmantown dolostone, Chazy and Black River-Trenton limestones. Hatch Point, at the very tip, is Canajoharie shale. The peninsula is separated from the mainland by Willsboro Bay, eroded in a suspected graben, with Precambrian marbles and anorthosite-related rocks on the mainland shore. The sharp break with mountains immediately west of Willsboro marks yet another major north-south block fault, with the west side upthrown.

Between Willsboro and Keeseville (12 miles), the road goes over this rugged Precambrian terrane, through Adirondack anorthosite and related rocks that are poorly exposed in old roadcuts (see Whiteface Mountain Memorial Highway discussion). This is the easternmost extension of the anorthosite body, which expands westward and southwestward and forms the backbone of all of the high peaks. Keeseville is at the northeastern boundary of this huge mass where it is overlain by Potsdam sandstone. The decidedly flat surface of the Champlain basin to the north reflects this change in bedrock foundation.

Canajoharie shale at Hatch Point; bedding dips gently to right, prominent jointing dips steeply to left. Courtesy Kendall / Hunt Publishing Co.

VI

SOUTHEASTERN NEW YORK, SHAWANGUNKS, CATSKILLS

The Port Jervis trough on the back slope of the Shawangunk Mountains is a major topographic feature that stretches for 100 miles along the base of the eastern scarp of the Allegheny Plateau from Stroudsburg, Pennsylvania, to Kingston, New York. On its east side, the backbone of the Shawangunk Mountains is the resistant Shawangunk conglomerate. Silurian beds rest on the so-called Taconic unconformity, an erosional surface on the underlying Ordovician shales, siltstones, and graywackes that were crumpled during the Taconian mountain-building event in late Ordovician time, 445-435 million years ago. The Silurian rocks were tilted, folded, faulted, and uplifted to their present position in the late Acadian uplift between 375 and 335 million years ago, along with further deformation of the rocks below the unconformity.

Also deformed were the early Devonian Helderberg/Onondaga group limestones and Hamilton group shales and siltstones that overlie the Silurian strata. The Acadian deformation, uplift, metamorphism, and extensive igneous activity were centered in New England to the east. Here, the Silurian beds were variably, though generally rather steeply, tilted to the northwest, faulted, and folded into a broad, open upfold, or anticline, that incorporates numerous smaller upfolds and downfolds, all with a northeasterly trend. The anticlinal character is only apparent in the northeastern section of the Shawangunk Mountains, between Ellenville and Stone Ridge, where it is broadly exposed over a width of about 6 miles. Southwest of Ellenville, the top of the anticline has been lost to erosion, leaving only its northwest limb as a narrow hogback, a term applied to a rather sharp-crested ridge held up by a steeply dipping resistant layer. These northwest-dipping beds everywhere form the east wall of the Port Jervis trough. The New Jersey extension of the Shawangunk Ridge is called Kittatinny Mountain.

The Catskills are carved from nearly flat-lying sedimentary rock strata of the thick eastern end of the Devonian Catskill Delta. Therefore, the topography here does not reflect differences in bedrock as in the Adirondacks; rather it results from differential erosion of the same bedrock.

Catskill Delta is the name applied to the enormous pile of middle and late Devonian sedimentary rock strata that underlies the entire Allegheny Plateau of New York and extends beyond the state to the south and west (see Introduction to the Southern Tier Expressway). The original sediments were generated by erosion of the Acadian Mountains as they came up in the New England region to the east. The sediments washed off the rising mountains into a shallow inland sea, gradually filling it and displacing the shoreline westward. Actually, the delta is a composite of coalescing alluvial fans and deltas formed by numerous streams that drained the western slopes of the mountains. Most of the delta is shale, but there are a few thin limestones, particularly near the base of the stratigraphic section, and coaser-grained sandstones and conglomerates are common in the upper part of the section.

The thickest rock section of the Catskill Delta is about 7500 feet in the Catskills, where reddish and greenish conglomerates and sandstones are commonplace. Post-Acadian erosion there, however, has been most extensive; and the original section may have been twice as thick. The delta thins westward to Lake Erie, as the sediments also get finer, reflecting greater distance from their source. The formations on the surface also get younger westward; the uppermost, or latest Devonian strata occur only in the western part. Taken as a whole, the

118

delta in New York, with the underlying early Devonian formations, has yielded one of the most complete fossil records of the Devonian period found anywhere in the world.

Topographically, the Catskills are one-sided mountains. The east side is a tremendous scarp, called the Catskill Mural Front, that faces the Hudson River for 30 miles and rises more than 2000 feet above it. Originally, the delta continued eastward to the Acadian Mountain slopes, across what are now the Hudson Lowlands and Taconic Mountains.

There is really no western side of the mountains, for there they merge gradually into the lower hill country that typifies the rest of the Allegheny Plateau. To the north, the mountains terminate in a kind of stair-step arrangement at the northern limit of the resistant middle to late Devonian Genesee group sandstones and conglomerates. The lower "tread," a thousand or more feet below, extends northward over middle Devonian Hamilton group beds to the Helderberg scarp, including the Helderberg Mountains in the northeastern corner of the plateau. To the south, the mountains also grade imperceptibly into the plateau.

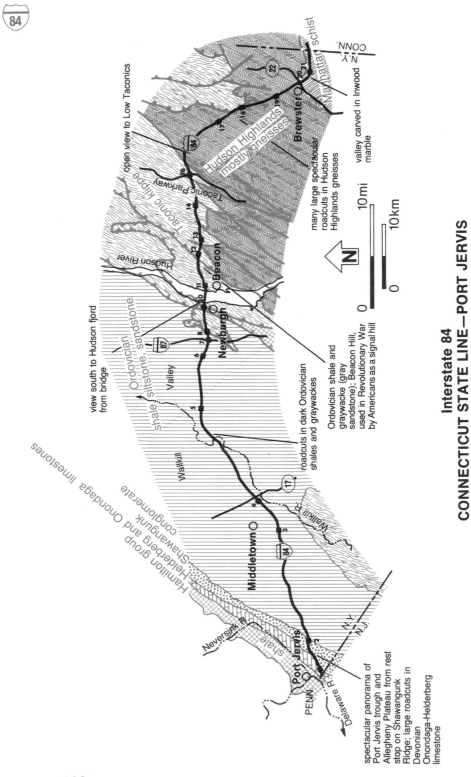

Interstate 84
CONNECTICUT STATE LINE—PORT JERVIS

open view to Low Taconics

Taconic klippe

Taconic Parkway

Hudson River

Hudson Highlands (mostly gneisses)

Manhattan schist

CONN.
N.Y.

Brewster

valley carved in Inwood marble

many large spectacular roadcuts in Hudson Highlands gneisses

Beacon

Ordovician shale and graywacke (gray sandstone); Beacon Hill, used in Revolutionary War by Americans as a signal hill

Newburgh

view south to Hudson fjord from bridge

Ordovician siltstone, sandstone

Valley

shale

roadcuts in dark Ordovician shales and graywackes

Wallkill

Wallkill R.

Hamilton group and Onondaga limestones

Shawangunk conglomerate

Helderberg and Onondaga limestone

Middletown

Neversink R.

shale

Port Jervis

Delaware R.

PENN.
N.Y.
N.J.

spectacular panorama of Port Jervis trough and Allegheny Plateau from rest stop on Shawangunk Ridge; large roadcuts in Devonian Onondaga-Helderberg limestone

N

10mi
10km
0
0

Interstate 84:
Connecticut State Line—Port Jervis
71 mi./115 km.

This is an east-west traverse across the structural grain of the northern end of the Hudson Highlands, southern end of the Taconic Klippe, and the Ordovician lowlands of the Wallkill Valley to the Shawangunk Ridge. It affords an excellent cross-sectional view of this geologically and physiographically complex region of New York State in a relatively short distance.

Between the Connecticut line and Brewster (4 miles), the road follows a valley developed in weak Inwood marble along the fault contact between the Manhattan Prong on the south and dark amphibolite, of the Hudson Highlands on the north.

Between Brewster and the Taconic State Parkway junction, (Exit 16, 15 miles), the highway crosses the Hudson Highlands in a north-westerly direction through numerous, large, new roadcuts in highly deformed gneisses and amphibolites. Several cuts display spectacular patterns of folding, faulting, and fluid flow. The Hudson Highlands, along with the Fordham gneiss and Adirondacks, contain the oldest rocks of New York State, formed in the core of the Ancestral Adirondacks during the Grenville mountain-building event, 1100-1300 million years ago (see also NY 22: White Plains—Berlin, Interstate 87: Tarrytown—Albany, and Taconic State Parkway discussions). The mountain range rimmed a coastline far inland from the present one, from Labrador to Mexico.

Topographically, this is a region of knobby, small, almost conical, hills lacking any conspicuous alignment, and lots of marshy, shallow valleys, small lakes, and large man-made reservoirs. In the last three miles east of Exit 16, the road descends about 700 vertical feet across a fault-line scarp at the Highlands boundary, with lovely, open views to the west over lowland of the frontal apron of the Taconic Klippe or low Taconics. This is the northern extension of the Ramapo fault through the Highlands; the valley immediately to the south of the highway has been eroded along it. Note here the sudden change in the roadcuts from the banded Highlands gneisses to dark mica schists and phyllites.

Between Exits 16 and 11 to Beacon (10½ miles), the highway crosses an exceedingly complex array of fault slices, along the chopped-up juncture between the Highlands to the south and Taconic Klippe to the north. The faults cross the highway in a northeasterly direction and include both block faults, and the thrust faults that are a trademark of the Taconics.

Between Exits 16 and 14 (15 is missing), cuts expose light gray Ordovician Wappinger group dolostones in nearly flat-lying, beds. Between Exits 14 and 13, the highway climbs over a ridge of quartz-plagioclase gneiss and quartzite in an upthrown block of Hudson Highlands that projects into the Taconic terrane. Exit 13 is back in Wappinger rocks. Between Exits 13 and 11, you cross three more major faults, with Highlands granitic gneiss in between. The last of these faults marks the western limit of the Highlands, and roadcuts near Beacon expose dark graywackes and shales of the Austin Glen formation, that are highly contorted and weathered to a rusty color.

As you cross the Hudson River on the Beacon-Newburgh Bridge, note the view south to the northern gateway to the Hudson fjord, with Storm King Mountain on the right and Beacon Hill-Breakneck Ridge on the left. In this scene, the importance of the Hudson Highlands as a barrier during the American Revolution can well be appreciated. At a time when there were virtually no roads through the mountains, the river constituted the only easy passageway; whoever controlled the gorge pretty much controlled the war. Save for a temporary breakthrough by three ships in 1777 which led to the burning of Kingston, the Americans held the Hudson fjord secure. They were thus able to contain the British occupation forces in New York City and Westchester County. They had a warning system of signal fires, or beacons, that stretched all the way from the Palisades to Albany. Each beacon site was on a high point visible from the next. One of these was on Beacon Hill behind the city of Beacon.

Newburgh lies in a small outlying erosional remnant of the Taconic Klippe on the west side of the river. Between Exits 10 and 8, at the northern edge of the city, there are roadcuts in buff-colored Wappinger dolostones with gently east-dipping beds and some minor faulting and folding. Between Newburgh and Port Jervis (34 miles), Interstate 84 traverses the Wallkill Valley, excavated almost entirely in weak Ordovician strata of the Normanskill and Martinsburg groups. Several roadcuts along the way expose dark gray to brown shales and graywackes of these units, many with gently west-dipping and slightly folded beds, and locally with strong cleavage that transects the bedding obliquely.

The highway crosses the Wallkill River near Montgomery between Exits 5 and 4. The river heads up in the Hudson Highlands and Kittatinny Mountain of New Jersey and flows northward all the way to Kingston where is joins the Hudson in a barbed junction that points northward against the flow of the master river. The Wallkill Valley is a remarkably broad lowland about 20 miles wide that stretches for 65 miles in a northeasterly direction. It is flanked on the east by the Hudson Highlands and Marlboro Mountain and on the west by the Shawangunk Ridge. The Wallkill River and tributaries are virtually alone in draining this huge region, save for the much shorter Moodna Creek in the central part, which enters the Hudson south of Newburgh. Obviously the Wallkill is much too small a river to have carved such a large depression by itself; rather, its geomorphic evolution is deeply rooted in Wisconsin glacial and deglacial events. There apparently was a pre-Wisconsin divide near Walden-Montgomery, with the ancestral Wallkill flowing northward from it, and another drainage system flowing southward. Glacial advance conveniently channeled through these well-developed valleys, gouged them wider and deeper, and removed the drainage divide. Meanwhile some of the same ice was rammed through the narrow Storm King gateway to the Hudson fjord. The southernmost stand of the Wisconsin ice sheet in the Wallkill Valley was in its headwaters region, where the Ogdensburg-Culvers Gap terminal moraine was deposited. Recession downvalley to the north began about 15,000 years ago, with several more moraines formed during temporary pauses. It was during this period that the valley served as a basin for a succession of meltwater lakes, including Lake Albany, banked up against the ice front, moraine, and high ground. Sediment deposition during this long episode further leveled the valley floor.

The highway intersects the Silurian boundary at the base of the Shawangunk Ridge about 5 miles east of Port Jervis near Exit 2. It then climbs over the ridge, which here is much less prominent than it is farther north near New Paltz. The height of the road, at 254 feet, is reached by a rest stop one mile west of Exit 2, with one of the grandest panoramas of the entire state. The view overlooks the spectacular

Delaware and Neversink valleys, 800 feet below, with the Allegheny Plateau of Pennsylvania and New York beyond. Port Jervis is at the point where the Delaware, flowing southeastward along the New York-Pennsylvania line, emerges from the plateau and makes a sharp right angle bend to the southwest, where it marks the Pennsylvania-New Jersey line. The lesser Neversink follows a similar trend a few miles to the north, joining the Delaware at Port Jervis. The contiguous northeast-trending valley of these two rivers, called the Port Jervis trough, is a major topographic feature that extends continuously for about 100 miles from Kingston, New York, where it intersects the Hudson Valley, southwestward to Stroudsburg, Pennsylvania, where the Delaware cuts across the Kittatiny Mountain ridge to form the famous Delaware Water Gap. All rivers from the west make similar sharp bends into the trough. Bedrock control of drainage and topograpy is obvious, with the valley excavated mostly in weak Hamilton group shales and siltstones that overlie the Onondaga and Helderberg limestones at the base of the Devonian stratigraphic section. The resistant Shawangunk conglomerate and basal Devonian units dip rather steeply northwestward here, so the trough is on the dip slope side of them.

US 209:
Port Jervis—Kingston
60 mi./97 km.

THE PORT JERVIS TROUGH

The entire route follows the Port Jervis trough on the backside of the Shawangunk Range. Between Port Jervis and Wurtsboro (19 miles), the valley is broad and open, with the mountains rising, in full view, 600-800 feet above either side. The trough has been excavated by stream and glacial erosion of the weak middle Devonian Hamilton group shales and siltstones of the backside of the Shawangunk Mountains. All the major streams that enter the trough from the Allegheny Plateau turn sharply and follow it for many miles. The Delaware, for example, has cut a deep gorge in the edge of the plateau northwest of Port Jervis. There it makes an abrupt right angle turn to the southwest to follow the trough all the way to Stroudsburg. At Stroudsburg, it again turns abruptly to the southeast and cuts across the Kittatiny Mountain ridge, an extension of the Shawangunk Ridge, in the famous Delaware Water Gap.

The valley floor is flat because, like so many major valleys of the glaciated regions of New York, it is filled with glacial outwash and lake sediments. The now-abandoned Delaware and Hudson Canal along its floor once provided a connecting link for barge traffic between these two rivers. The flatness, and steepsidedness are stikingly apparent from high, breathtaking, panoramic viewpoints on cross-highways, such as Interstate 84 east of Port Jervis, or NY 17 east or

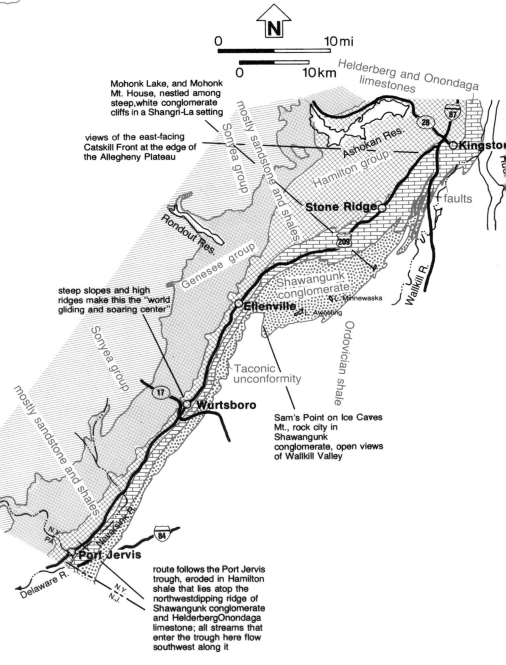

209

0   10 mi

0   10 km

Helderberg and Onondaga limestones

Mohonk Lake, and Mohonk Mt. House, nestled among steep, white conglomerate cliffs in a Shangri-La setting

views of the east-facing Catskill Front at the edge of the Allegheny Plateau

mostly sandstone and shales

Sonyea group

Ashokan Res.

Hamilton group

Kingston

Stone Ridge

faults

Genesee group

Rondout Res.

209

Shawangunk conglomerate

L. Minnewaska
L. Awosting

Wallkill R.

steep slopes and high ridges make this the "world gliding and soaring center"

Ellenville

Ordovician shale

Sonyea group

Taconic unconformity

17

Wurtsboro

Sam's Point on Ice Caves Mt., rock city in Shawangunk conglomerate, open views of Wallkill Valley

mostly sandstone and shales

84

N.Y.
PA.

Port Jervis

Delaware R.

N.Y.
N.J.

route follows the Port Jervis trough, eroded in Hamilton shale that lies atop the northwestdipping ridge of Shawangunk conglomerate and HelderbergOnondaga limestone; all streams that enter the trough here flow southwest along it

US 209
PORT JERVIS—KINGSTON

west of Wurtsboro. Wurtsboro, incidentally, touts itself as the "World gliding and soaring center," a tribute to the suitable character of the surrounding topography.

Between Wurtsboro and Ellenville (13 miles), there is a "through-valley" drainage divide and narrow constriction in the valley. Streams on one side of the divide, at Phillipsport, flow southwestward; and, on the other side, northeast-flowing small creeks form the headwaters of Rondout Creek. This divide is apparently the site of a Tertiary bedrock barrier that was reduced by glacial erosion. It is enhanced by a pitted moraine that marks a minor stand in the recession of the Wisconsin ice sheet.

Dark brown, Hamilton group siltstones and shales are exposed in roadcuts near Ellenville. A well-defined rock terrace here in the Hamilton beds west of the village, is elevated about 350 feet above the valley floor. The "Gunks" here are much higher than they are near Port Jervis, rising a maximum of about 1900 feet above Ellenville.

Route 52, east of Ellenville, goes to Ice Caves Mountain at the crest of the "Gunks." The mountain is at the southernmost tip of the broad, anticlinal section of the ridge where flat-lying, erosionally truncated beds of white Shawangunk conglomerate form a cliff over the Taconic unconformity. The dark shales of roadcuts en route from Ellenville are Ordovician beds below the unconformity. The so-called Ice Caves part of the cliff is really a "rock city" like those that occur in the late Devonian, Mississippian, and Pennsylvanian conglomerates of southwestern New York (see NY 17: Arkport/Hornell—Westfield discussion). The massively-bedded, widely-jointed rock at the edge of the cliff breaks up into huge blocks that tumble and lean against each other as they gradually work their way downslope over the shales. The term "city" comes from the resemblance of the rectangular blocks to buildings, and of the passageways to streets. The coldness of "streets" and caves probably stems from the long-lasting winter ice trapped in them.

Sam's Point, at the southern limit of the Shawangunk cliff, offers a panorama that encompasses the Wallkill Valley, the ridge to the southwest, the Port Jervis trough, and the Allegheny Plateau. The view of the scarp to the northeast is practically blocked by trees. The point is named for Samuel Gonzales, who, according to folklore, was chased to the edge of the cliff by a band of Indians during the French and Indian War. Seeing no other escape, he jumped from the cliff into the hemlock boughs below and escaped unharmed.

Between Ellenville and Stone Ridge (18 miles), you follow Rondout Creek downstream. This section of the valley is less trough-like, and the creek meanders widely. A massive stone firetower called Skytop high up on the Shawangunk crest, is visible over part of the way.

Sam's Point on Shawangunk conglomerate

Skytop overlooks a "Shangri-la" of incredible beauty. You have to see it to believe it. Nestled among craggy cliffs of dazzling white conglomerate garnered with evergreens and hardwoods is crystal clear, deep, blue-green Mohonk Lake. At the other end of the lake is the massive Mohonk Mountain House with its unusual assortment of interconnected buildings of varied architectural styles and a seemingly endless array of roof peaks and turrets. Viewed from Skytop, all seems suspended in mid-air above the Rondout Valley/Port Jervis trough, over a thousand feet below, as if it had drifted away from the distant eastern scarp of the Allegheny Plateau.

Mohonk Lake is one of three gem-like, rock-walled lakes in this section of the mountains; the others are Lake Minnewaska and Lake Awosting, a few miles to the south. All three lake basins are thought to lie along fault zones, the broken rocks of which were scooped up and carried away by glaciers during the Ice Age. Glacial polish and scratches are abundantly preserved on the top surfaces of the durable quartz pebble conglomerate.

The "Gunks," as all mountain climbers know, constitute the best rock-climbing region of the eastern United States. The rock is unusually solid and trustworthy; and the bedding in most cliffs dips into the hill, giving excellent handholds, footholds, ledges, and overhangs. Vertical joints, meanwhile, offer a great variety of crack and chimney climbs; and the steep faces make spectacular rappels. Unfortunately, the climbs, practically all of which are described in guidebooks, are now so popular that one often has to wait in line to do them.

Mohonk Lake and the Mohonk Mountain House from Skytop overlook; white cliffs of Shawangunk conglomerate

Between Stone Ridge and Kingston (11 miles), the road continues along a section of the trough floored by the Onondaga limestone, exposed in several roadcuts. Rondout Creek cuts across the Shawangunk Ridge near Stone Ridge, then goes north to Kingston, where it joins the Hudson. The ridge itself quickly dies out north of Stone Ridge as the conglomerates that support it and, in fact, all of the Silurian strata diasappear. The prominent scarp north of Kingston is the Catskill Mural Front at the eastern edge of the Allegheny Plateau.

THE CATSKILLS—INDIVIDUAL GEOLOGIC SITES

1. Catskill Mountain House

This magnificent viewpoint is at the rim of the Catskill Mural Front, the eastern scarp of the Catskill Mountains, 2100 feet above the Hudson River and only about 6 miles west of it. To reach the site, take the North Lake road from the village of Haines Falls on NY 23A. The road goes three miles east to a public campsite at North Lake, from which you can reach the overlook by an easy 5-minute walk.

The Catskill Mountain House is no longer there, but the scenery is intact. The great resort hotel stood at the very brink of the precipice from 1824 until 1963; by that time it had fallen into a sad state of disrepair, and the New York State Department of Conservation, which had acquired its lands, burned it to the ground. In the early days, guests reached the hotel by stagecoach from the Catskill boat landing on the Hudson. After the Civil War a railroad was constructed from Kingston, with a spur leading to the Mountain House along approximately the same route as the present North Lake road. In 1892, an inclined railway was built up the face of the scarp to the hotel. These are gone now, and the land has been returned to its natural state; but the hotel, railroad grades, and even the hotel entry road are still shown on the 1980 edition of the Kaaterskill quadrangle of the U.S Geological Survey 7½ minute series topographic maps.

The ledge upon which the Mountain House stood is coarse-grained conglomeratic, reddish to greenish, cross-bedded sandstone belonging to the Oneonta formation of the late Devonian Genesee group. It

Catskill Mountain House ledge and the Catskill Mural Front; Hudson River to right and 2000 feet down

is part of the resistant caprock that holds up the scarp. More of the same may be seen along the rim to the north and south of the hotel site at higher elevation. The scarp is not a single steep slope, but a series of sandstone/conglomerate steps separated by shaley interbeds. Its exceptionally straight trend is the product of glacial abrasion by the Hudson lobe of the Wisconsin ice sheet. Glacial polish and striations are well-preserved on some sandstone surfaces. North and South lakes, back of the hotel site, occupy shallow basins scooped out by the passing ice.

The incredible view from here encompasses a 50-mile sweep of the Hudson Valley from south of Albany almost to the Hudson Highlands. There is, perhaps, no better place to gain an appreciation of the immensity of geologic events that have shaped the state. Imagine, for instance, the Catskill Delta piled another mile or so over your head and extending *upslope* to the high Acadian Mountains in New England—piled over the worn-down stumps of the earlier Taconic Mountains, the even-further-reduced remnants of which are now visible as low rumpled ridges beyond the Hudson. And then, imagine all of that titanic pile being eroded to the present landscape, and the sediments being carried away to form a new pile somewhere else. Makes one feel very small and insignificant, doesn't it?

2. Kaaterskill Clove—The Rip van Winkle Trail

"Clove" is the name used for most of the deep, postglacial ravines of the Catskills. Kaaterskill Clove is the largest of two accessible cloves in the Catskill Mural Front, the other being Plattekill Clove, about 4 miles to the south. In the 5 miles from Palenville to Haines Falls, NY 23A, the Rip van Winkle Trail, ascends 1400 feet in a nearly straight line from the base of the clove to the top. Near the very steep road there are many excellent bedrock exposures, mostly along the stream bed. The lower half of the clove cuts through Hamilton group red shales with sandy interbeds. The upper half is in the Oneonta formation and contains more and thicker resistant sandstone beds separated by recessively weathering red shales. The many falls en route are held up by the sandstone.

3. Plattekill Clove

If anything, Plattekill Clove is even more awesome than Kaaterskill, dropping the same vertical distance in just 3 miles, instead of five. The same formations are exposed as in Kaaterskill Clove. The road is very steep, so gear down: The road is the Platte Clove road between the hamlets of West Saugerties and Platte Clove.

Plattekill Clove, a slice into the Catskill Mural Front

4. Kaaterskill Falls

A short but rugged trail from the Kaaterskill Clove road takes you to the falls. The trail begins at the only hairpin turn in the road, where it crosses the main branch of Kaaterskill Creek coming from the falls; there is a parking lot nearby. The falls cascade over two sandstone steps very near the Hamilton-Oneonta contact. The supporting shale under each ledge has weathered back, leaving the sandstone overhanging. The upper falls drop 180 feet and the lower ones 80 feet so the combined drop of 260 feet is the highest of any cascade in New York. You reach the top of the falls from the North Lake road, but the view and geology from there are less impressive.

5. Folded and Faulted Helderberg Limestones

These very large roadcuts are on NY 23A a short distance west of the Interstate 87 intersection. They illustrate Acadian deformation of the early Devonian Helderberg limestones that underlie the Catskill Delta sediments. Exposed here on both sides of the highway is a beautiful upfold, or anticline, broken by low-angle thrust and bedding faults, along which beds have been drag-folded. A lot of the limestone has been brecciated—broken up—by the deformation.

132

Kaaterskill Falls, in two sandstone steps, near the Hamilton/Oneonta contact (Devonian)

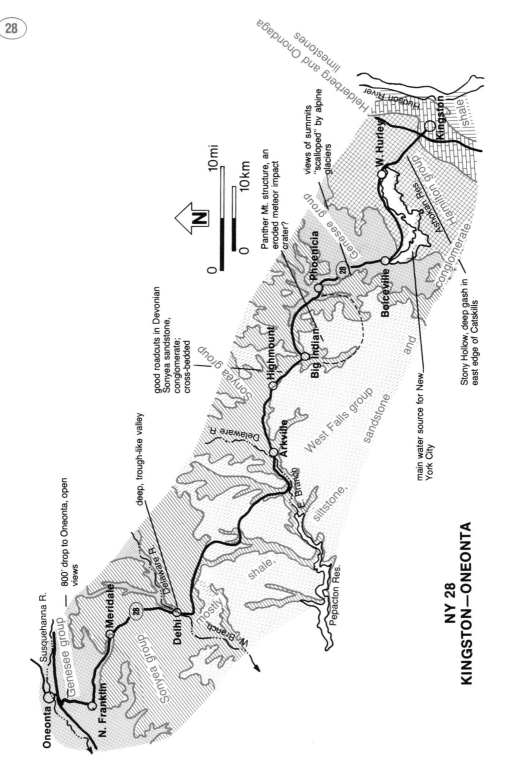

Helderberg and Onondaga limestones

Hudson River

shale

Kingston

W. Hurley

Hamilton group

Ashokan Res.

conglomerate

views of summits "scalloped" by alpine glaciers

Genesee group

Panther Mt. structure, an eroded meteor impact crater?

Phoenicia

28

Boiceville

Stony Hollow, deep gash in east edge of Catskills

good roadcuts in Devonian Sonyea sandstone, conglomerate; cross-bedded

Sonyea group

Highmount

Big Indian

West Falls group

sandstone

main water source for New York City

Arkville

Delaware R.

E. Branch

siltstone,

Pepacton Res.

deep, trough-like valley

shale,

mostly

W. Branch

Delaware R.

800' drop to Oneonta, open views

Susquehanna R.

Genesee group

Meridale

28

Delhi

Sonyea group

Oneonta

N. Franklin

0   10 mi

N

0   10 km

NY 28
KINGSTON—ONEONTA

134

NY 28:
Kingston—Oneonta
90 mi./146 km.

Between Kingston and West Hurley (6 miles), NY 28 passes through Stony Hollow, one of the deep gashes carved by small streams in the eastern edge of the Catskills. The several large road-cuts en route expose flat-lying beds of brownish to reddish-brown shales and sandstones of the middle Devonian Hamilton group. The high peaks of the Catskill Mountains proper lie still farther west beyond the high scarp of the Catskill Mural Front. The Hamilton beds here form an intermediate step to the mountains. The north-south, vertical jointing so prevalent in some cuts is a regional feature on which a rectangular drainage pattern of small streams has developed.

From West Hurley to Boiceville (10 miles), you follow the north shore of Ashokan Reservoir, the main source of water for New York City. From here, water is piped through the 77-mile Catskill aqueduct to the Kensico holding reservoir north of White Plains, on the way passing under the Hudson River at Storm King Mountain (see NY 22: White Plains—Berlin discussion).

Ashokan Reservoir is formed by damming of Esopus Creek, with the main dam on the south side about midway in the length of the lake. Downstream the Esopus empties to Rondout Creek which, in turn, flows northeastward to the Hudson at Kingston. There are excellent open views around the lakeshore to the high mountains, which rise immediately west of the lake.

Between Boiceville and Phoenicia (8 miles), you follow the Esopus upstream. Phoenicia calls itself the "tubing capital of the world," where thousands of fun-seekers ride the rapids of the Esopus on innertubes each summer. This section of highway is especially good for seeing the "scalloping" of the high peaks, a product of "alpine" glacial erosion preceding and/or following the major ice sheet invasions of the Ice Age. Small glaciers that coursed down preglacial stream valleys carved the bowl-shaped cirques, or amphitheaters, and their U-shaped valleys below (see Whiteface Mountain Memorial Highway discussion). These features are much better developed here than in the Adirondacks probably because the rock strata are flat-lying, and there is a lot of weaker rock under the sandstone/conglomerate caprock that could be more easily hollowed out.

"Scalloped" hill near Phoenicia, hollowed out by alpine glacial erosion

The route between Phoenicia and Big Indian (9 miles), continues upstream along Esopus Creek, following the valley rim of the so-called "Panther Mountain Circular Structure." Esopus Creek continues encircling to the south up Big Indian Hollow for another 8 miles to its headwaters in the south-central part of the rim. The eastern and southeastern parts of the rim are delineated by Woodland Creek, which joins the Esopus just west of Phoenicia and has headwaters narrowly separated from those of the Esopus by the Panther Mountain ridge. The nearly complete and nearly perfect circle formed by the valley has a diameter of about 6 miles. Recent geological, geophysical, and satellite-imagery studies suggest that this unusual feature may well be the erosional expression of a meteorite impact crater formed in late Devonian time, and subsequently buried under sedimentary strata. Under this interpretation, the valley defines the crater rim; and it has been preferentially eroded because the cover rock there was more highly fractured than elsewhere.

Between Big Indian and Highmount (4 miles), at the top of the pass by the Belle Ayr Mountain ski area and at the western border of the Catskill Forest Preserve, the road climbs 700 feet up Birch Creek valley. Roadcuts are in grayish to brownish, crossbedded sandstones and conglomerates of the late Devonian Sonyea group. Crossbedding is a common feature of the coarser-grained rocks of the Catskills that indicates deltaic, near-shore deposition. The beds are inclined to the normal bedding and truncated by it at the top and bottom. In general, they represent the "foreset beds" of the building deltas, while the "topset beds" are the normal, horizontal beds. A lot of the Catskill rocks cleave easily into flagstone along the crossbedding planes.

From Highmount to Arkville (7 miles), the road descends 600 feet down Bush Creek valley to the east branch of the Delaware River. The roadcuts are similar to those of the other side of the pass. From here, the erratic course of the Delaware goes southwestward to Hancock, Pennsylvania, where it makes a sharp-pointed junction with the main branch that flows southeastward along the border between the two states (see NY 17: Harriman—Binghamton discussion).

Between Arkville and Delhi (25 miles), you first follow the Delaware downstream to the Pepacton Reservoir, then go overland, crossing the divide to the Little Delaware River, then follow it downstream to its junction with the west branch of the Delaware at Delhi. Several of the higher roadcuts are in reddish sandstones and shales of the late Devonian West Falls group that overlies the Sonyea group. Many points en route provide open views of the attractive rolling landscape characteristic of the plateau. The plateau label is somewhat misleading; the image it conjures of a flat surface is never realized. However, it was more plateau-like before it was so deeply dissected by streams and ice. Stream dissection, of course, still goes on today. The pronounced rounding of the hilltops is a product of overriding by ice.

Delhi is attractively nestled in the narrow Delaware Valley just north of the Little Delaware juncture. The west branch of the river from here follows a course nearly identical to that of the east branch, flowing southwestward to join the main branch in a "barbed junction" near the Pennsylvania border. Barbed junctions often indicate stream "piracy," whereby headwater extension of one stream intercepts the course of another and steals its upstream water. The big question in this region is whether the Delaware pirated the Susquehanna, or vice-vesa.

The valley at Delhi exhibits the glacial trough characteristics of steep walls, truncated spurs and a flat, sediment-filled floor. Between Delhi and Meridale (8 miles), you go overland to Ouleout Creek and then follow it downstream to North Franklin (8 miles). The Ouleout Valley is remarkably asymmetrical with a steep south wall and gentle north wall. Many east-west trending valleys of the glaciated plateau developed similar asymmetry as a result of southward overriding by glacial ice. In the process, till was smeared on the south-facing slope in much the same way that sand, driven by wind or water, accumulates on the lee sides of obstacles. These are referred to as "till-shadow hills."

Between North Franklin and Oneonta (5 miles), you cross the divide to the Susquehanna Valley, with open views on the steep 800-foot descent to the city, both of the valley itself and of the plateau to the north (see Interstate 88: Schenectady—Binghamton discussion).

NY 28:
Oneonta—Mohawk
50 mi./81 km.

This route follows the Susquehanna River for 23 miles between Oneonta and Cooperstown at the south end of Otsego Lake, one of the principal sources of the river. En route, you pass Charlotte Creek just east of Oneonta, Schenevus Creek at Colliersville, and Cherry Creek at Milford, all of which enter the Susquehanna from the northeast. Oaks Creek, from Canadarago Lake, enters from the northwest just south of Cooperstown. The section of valley between Oneonta and Colliersville is discussed in Interstate 88: Schenectady—Binghamton.

North of Colliersville, the valley closely resembles the downstream section; it is broad and flat-floored, with abrupt, steep walls and truncated spurs, a lot of pitted outwash, and well-developed sand/gravel terraces along the valley sides. The origins of these features are discussed in some detail in the Interstate 88 section. Basically this, and all of the tributaries mentioned above, are glacial troughs formed in preglacial stream valleys that were deepened and widened by glacial erosion, and then partially filled with glacial drift during Wisconsin deglaciation. Otsego and Canadarago lakes are finger lakes formed in sections of the troughs that are plugged by moraines at their southern ends. Unlike the Finger Lakes to the west, however, these lakes drain southward instead of northward because they have breached their moraine dams. The moraine is called the Cassville-Cooperstown moraine; it crosses the Susquehanna Valley about 2

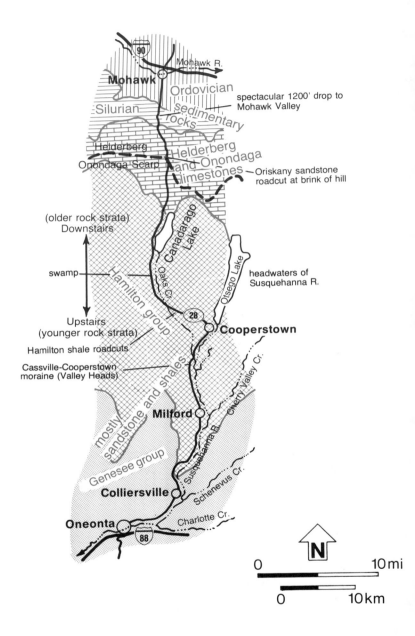

Mohawk R.

Mohawk

90

Ordovician

Silurian

sedimentary

rocks

spectacular 1200' drop to
Mohawk Valley

Helderberg

Onondaga Scarp

Helderberg
and Onondaga
limestones

Oriskany sandstone
roadcut at brink of hill

(older rock strata)
Downstairs

Canadarago
Lake

swamp

Hamilton group

Oaks Cr.

Otsego Lake

headwaters of
Susquehanna R.

Upstairs
(younger rock strata)

28

Cooperstown

Hamilton shale roadcuts

Cassville-Cooperstown
moraine (Valley Heads)

mostly
sandstone and shales

Cherry Valley Cr.

Milford

Genesee group

Susquehanna R.

Schenevus Cr.

Colliersville

Oneonta

88

Charlotte Cr.

N

0   10 mi

0   10 km

**NY 28
ONEONTA—MOHAWK**

140

miles south of Cooperstown. Erosional remnants continue for some distance up Oaks Creek valley. Other patches of moraine in Red Creek and Cherry valleys to the east represent the same ice stand.

Cooperstown is, of course, known best for its National Baseball Hall of Fame. Between Cooperstown and Fly Creek (3 miles), the highway crosses the spur that separates the Otsego basin from Oaks Creek. The gray shales and sandstones of the roadcuts belong to the middle Devonian Hamilton group. Note that the traverse from the Catskills northward along NY 28 has crossed progressively older Devonian beds.

Between Fly Creek and Richfield Springs (12 miles), you follow Oaks Creek to its source at Canadarago Lake, then, with open views, traverse the west shore of the lake to its northern end. The hummocky moraine topography around Fly Creek continues upvalley for about 2 miles, where it ends abruptly against flat, marshy lake plain that extends north to the lake. Practically all of the finger lakes have these; they were formed during Wisconsin deglaciation when the lake levels were higher and banked against the moraine.

Numerous drumlins, with elongation and asymmetry indicating an unusual east-to-west ice advance, dot the plateau north of the lakes. Their origin is similar to that of the Duanesburg drumlin field discussed in Interstate 88: Schenectady—Binghamton.

The road between Richfield Springs and Mohawk (12 miles) first crosses the drumlin field to the northern rim of the plateau and then makes a spectacular headlong plunge into the Mohawk Valley in the last 2 miles, dropping a total of over 1200 feet. There is an excellent cut in brownish early Devonian Oriskany sandstone at the brink of the hill. The descent requires the driver's full attention. If you were to walk down, however, you would have some marvelous views over the Mohawk Lowlands and Helderberg/Onondaga scarp east and west.

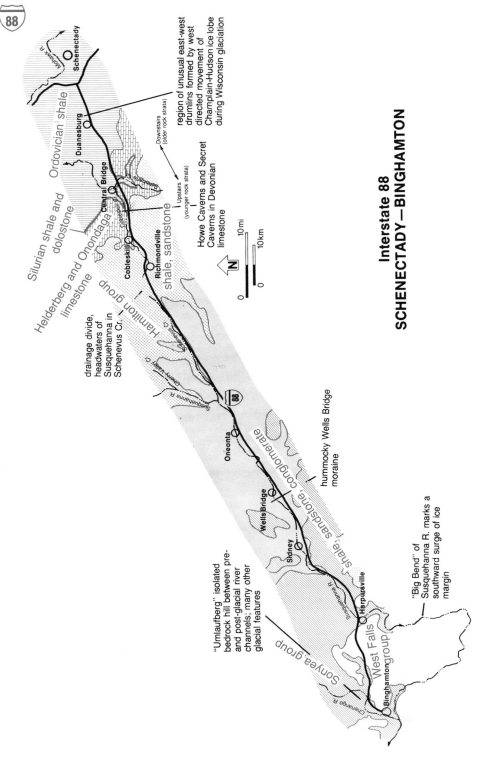

Interstate 88
SCHENECTADY–BINGHAMTON

region of unusual east-west drumlins formed by west directed movement of Champlain-Hudson ice lobe during Wisconsin glaciation

Downstairs (older rock strata)

Upstairs (younger rock strata)

Howe Caverns and Secret Caverns in Devonian limestone

Ordovician shale

Silurian shale and dolostone

Helderberg and Onondaga limestone

Hamilton group shale, sandstone

drainage divide, headwaters of Susquehanna in Schenevus Cr.

hummocky Wells Bridge moraine

"Umlaufberg" isolated bedrock hill between pre- and post-glacial river channels; many other glacial features

"Big Bend" of Susquehanna R. marks a southward surge of ice margin

shale, sandstone, conglomerate

Sonyea group

Binghamton group

Schenectady

Mohawk R.

Duanesburg

Central Bridge

Cobleskill

Richmondville

Cobleskill Cr.

Schenevus Cr.

Chem. Valley Cr.

Susquehanna R.

Oneonta

WellsBridge

Sidney

Harpursville

West Falls

Binghamton

Chenango R.

Susquehanna R.

N

10mi

10km

Interstate 88:
Schenectady—Binghamton
120 mi./194 km.

Between Schenectady and Central Bridge (22 miles), Interstate 88 crosses the Mohawk Lowlands over middle-late Ordovician Schenectady formation graywackes, sandstones, siltstones, and shales. Near Duanesburg, you cross a dense field of drumlins, representing one of the few places in New York where Wisconsin ice moved from east to west. The glacial history recorded in the Duanesburg region is as follows. During Wisconsin glaciation, when the southward-advancing Hudson-Champlain ice lobe first reached the broad expanse of lowland near Albany, the ice spread out and a lobe began moving up the Mohawk Valley, guided by the east-west trending Helderberg scarp at the northern margin of the Allegheny Plateau. The lobe continued its westward migration all the way to Herkimer, where it met head-on with the eastward-advancing Oneida lobe in a kind of pincer action around the Adirondacks. Confinement of the ice to the Mohawk Lowlands, however, was only temporary until its thickness had built up sufficiently to mount the Allegheny Plateau and move southward over it. The net result is the unusual juxtaposition of east-west drumlins at Duanesburg, and north-south drumlins a few miles southeast of them atop the plateau in the Helderberg Mountains.

The route between Duanesburg and Central Bridge (11 miles) goes high up on the slope with intermittent views north across the Mohawk Lowlands to the Adirondack foothills. It then drops to the

Schoharie Valley, with open views north and south. Several roadcuts en route are in Schenectady formation shales with sandstone/siltstone interbeds.

Central Bridge is at the junction of Cobleskill Creek with Schoharie Creek. With its headwaters in the Catskill Mountains, Schoharie Creek is the only major river in the eastern Allegheny Plateau that drains northward across the Helderberg scarp, joining the Mohawk River at Fort Hunter. The valley was extensively scoured during the Ice Age and sedimented during Wisconsin deglaciation. At one stage, a long, slender glacial lake called Lake Schoharie filled the valley south of a moraine dam at Esperance and extended arms up the Cobleskill and Fox valleys. That section is now especially broad and flat, as it is partially filled with lake and delta sediments.

Cobleskill, 11 miles up the Cobleskill Valley from Central Bridge, is cave country. Most of the caves, including the well-known Howe Caverns, are in the section of the Allegheny Plateau north of Cobleskill called the Cobleskill Plateau, which is capped by the early/middle Devonian Helderberg and Onondaga limestones. The geologic setting is ideal for cave development. Consequently, there are more caves here than anywhere else in the northeastern United States. A similar geologic environment occurs along Fox Creek, east of Schoharie Creek.

Howe Caverns, with flowstone and dripstone
Rodney Schaeffer photo,
courtesy Kendall/ Hunt Publishing Co.

Caves form when groundwater moves through openings in limestone, dolostone, or marble, and dissolves them over long periods of time. The ability of the water to dissolve the rock is directly related to its acidity, which depends on how much carbon dioxide it picks up from the air, or organic acid from the soil. This slightly acid water, moving along bedding planes, joints, and faults, is able to enlarge them into sinks, tunnels, passageways, and large chambers. To do so, the water must flow through the rock openings, continually carrying the dissolved carbonate away.

The bedrock strata of the Cobleskill Plateau dip just 1-2 degrees to the south-southwest. Water enters through "insurgents" in the upper part of the plateau fed by small streams tributary to Schoharie Creek. It then seeps southward down the dip of the beds and along joints and faults, hollowing them out, and emerges through "resurgents"—springs—in the Cobleskill Valley, or on its north side.

Most of the caves of this region appear to be preglacial in origin. Glaciation had little effect on the underground caves themselves, while it greatly modified the surface solutional topography, or "karst." Glacial ice crushed or quarried many of these features; therefore, the sinks and other karst phenomena exposed on the plateau today are probably postglacial. Thick deposits of till and drumlins on parts of the plateau now prevent insurgence of surface water and cause extensive derangement of surface drainage patterns. Many of the current sinkholes result from collapse of glacial sediments into underlying caves.

Cobleskill Valley has been filled with glacial debris to about 100 feet, displacing the present stream southward onto a limestone bench cut by the preglacial stream. Drumlins are fairly numerous on the

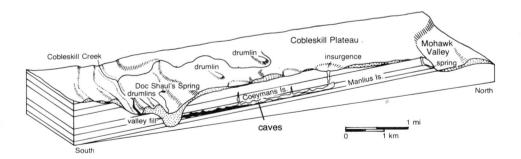

Cross-section of Cobleskill Plateau showing gently south-dipping limestones, caves, and glacial features   Courtesy Kendall / Hunt Publishing Co.

valley floor between Howes Cave village and Cobleskill, indicating a southwest ice advance. At one place on the north side of the valley, resurgent water from McFail's Cave has "piped" through the valley fill to form Doc Shaul's Spring. In preglacial, or at least pre-Wisconsin time, this was a gravity spring that issued directly from the limestone of the north slope of the valley.

The south wall of the Cobleskill Valley is extensively scoured by the ice. Most conspicuous is Ecker Hollow, a beautiful U-shaped hanging valley 500 feet above Cobleskill Creek about 4 miles east of Cobleskill.

The highway follows Cobleskill Creek upstream from Cobleskill to Richmondville (6 miles), where a barely perceptible divide separates this tiny stream from equally tiny Schenevus Creek. Schenevus Creek, in turn, flows southwestward from the divide to join the Susquehanna River. The Cobleskill, here, has not incised deeply enough to expose the cave-forming limestones, and thus resurgent water has no easy way out. Caves are less common.

Four miles west of Richmondville, the road passes through a narrow bedrock valley that probably is the site of a preglacial drainage divide reduced by ice and stream erosion. Roadcuts expose middle Devonian Hamilton group shales, siltstones, and sandstones.

Between Richmondville and Oneonta (30 miles), you follow Schenevus Creek downstream. It joins the Susquehanna River at Colliersville, 4 miles northeast of Oneonta. The city of Oneonta is at a point on the broad Susquehanna sediment plain where the river branches into several large tributaries, including Schenevus and Charlotte creeks from the northeast and Otsego Creek from the north. Cherry Valley Creek also joins the main branch a few miles to the north. The main valley floor here and downstream is choked with 200-300 feet of drift, mostly of glacial lake and till origin. In places, conspicuous terraces constructed of sandy, gravelly sediments mark the valley sides. These have been interpreted as "delta terraces" formed at the mouths of numerous small streams that flowed into a proglacial lake from the valley sides. Particularly good ones are southeast of Oneonta on the south side of the river.

Between Oneonta and Wells Bridge (13 miles), Interstate 88 follows the Susquehanna downstream. Much of the floodplain north and south of the highway was under water during the spring flood of 1977, when the highway itself was nearly submerged. In passing Otego, the road rises on the south wall, with good views over the modern floodplain and an abrupt change in the valley trend that is a remnant of a preglacial "engrown" meander. Engrown meanders are characteristic of many parts of the Susquehanna Valley. They result from the

146

downcutting of stationary meander loops. They were initiated by the ancestral Susquehanna in Tertiary time before deep dissection of the plateau began as a consequence of regional uplift.

Near Wells Bridge, you cross over the hummocky, pitted surface of the Wells Bridge moraine. This moraine completely blocked the up-stream side of the valley during glacial retreat, forming Lake Otego from here to Oneonta. The delta-terraces mentioned earlier were built along the shores of this narrow strip of a lake, and most of the valley fill under the modern floodplain deposits consists of Lake Otego sediments.

The valley floor downstream from the Wells Bridge moraine nearly to Sidney (10 miles) is outwash plain with a few small kettles where blocks of stagnant ice were buried in outwash. The road traces an-other large engrown meander around Sidney, and yet another at Bainbridge 5 miles farther west. Sidney is on a broad sediment plain at the Unadilla River junction. All the rivers here are "underfit" because their valleys were enlarged first by glacial erosion and sec-ondly, by the great volumes of meltwater that passed through them during glacial retreat. Nearly all of the bedrock valleys are deeply buried in drift and modern sediments.

Several roadcuts around Unadilla and Sidney in the Unadilla and Oneonta formations of the middle-late Devonian Genesee group ex-pose gray to brown, thin-bedded shales, siltstones, and sandstones, and even some redbeds.

The road continues to follow the Susquehanna between Bainbridge and Harpursville (12 miles). Pitted moraine-outwash occurs just east of Afton, with two large kettle lakes and several smaller ones. The section of valley near Harpursville is distinguished by extremely broad, flat terraces about 100 feet above the river, apparently formed as delta terraces marginal to a receding ice tongue.

Between Harpursville and Binghamton (19 miles), the road crosses the divide that separates the two sides of the Great Bend of the Susquehanna (see NY 17: Harriman—Binghamton for further dis-cussion). Roadcuts at the beginning and end of the traverse near the river expose gray to brown shales, siltstones, and sandstones of the late Devonian Sonyea group, while cuts on the upland expose the stratigraphically higher shales of the West Falls group. You descend to the Chenango River at Chenango Bridge and then follow it downstream for 5 miles to its junction with Susquehanna.

There is no better place in New York State to see stagnant ice features than the southern Chenango River valley. A hummocky, moraine-kame complex dominates at the north end of this stretch of road in Chenango Valley State Park near Chenango Forks. This gives way southward to pitted outwash with tens of small kettles and two large kettle lakes separated by a ridge of sand and gravel originally deposited as a crevasse filling. Pitted outwash continues south to Chenango Bridge. One of the most striking features is the rounded umlaufberg visible by Chenango Bridge where Interstate 88 descends to the river. This is a bedrock hill that was erosionally isolated by ice-marginal drainage changes. The preglacial river channel was north of the hill. An ice stand along the present course initiated a new channel while the old channel was still plugged. With further retreat, the river continued in the new channel where it remains today.

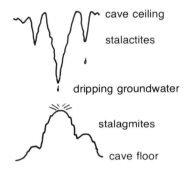

Formation of dripstone in caves. Each drop releases a small amount of $CaCO_3$ that builds up in time to the forms shown.

NY 12:
Binghamton—Utica
85 mi./138 km.

This is the route of the abandoned Chenango Canal, one of the most unsuccessful of the canals built during New York's canal era. The canal was built between 1834 and 1836, opened in 1837, and abandoned in 1878 at a time when the state was trying to rid itself of the financial burden of the non-profitable canals. This one cost the taxpayers dearly: the total cost of construction, improvements, and operation was nearly $7 million, while the total revenues collected in its 42 years of operation amounted to only about $750,000.

As in all of New York's canals, the Chenango followed natural routes as much as possible. It was, however, a "lateral canal" that cut across geological trends; and this, in all cases, meant that many locks were needed to overcome great differences in elevation from one end to the other. This one required 116 locks for a total distance of 97 miles, or an average of more than one lock per mile! No wonder it was costly. This, compare with the Genesee Valley Canal from Rochester to Olean, also a lateral canal with 112 locks in 124 miles (see discussion of Field Sites in Letchworth Gorge). The losses there were similar, but the Genesee Valley Canal only operated for 26 years. A constitutional amendment adopted in 1874 permitted the sale or abandonment of all canals owned by the state except the Erie, Champlain, Oswego, and Cayuga and Seneca. Those are the only ones that are still operating today.

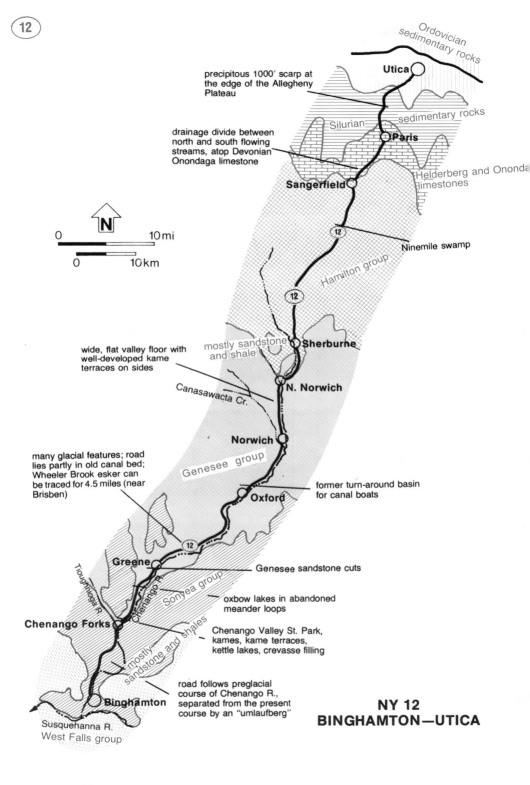

Ordovician sedimentary rocks

Utica

precipitous 1000' scarp at the edge of the Allegheny Plateau

sedimentary rocks

Silurian

Paris

drainage divide between north and south flowing streams, atop Devonian Onondaga limestone

Helderberg and Ononda limestones

Sangerfield

N

0   10 mi

0   10 km

12

Ninemile swamp

12

Hamilton group

wide, flat valley floor with well-developed kame terraces on sides

mostly sandstone and shale

Sherburne

N. Norwich

Canasawacta Cr.

Norwich

many glacial features; road lies partly in old canal bed; Wheeler Brook esker can be traced for 4.5 miles (near Brisben)

Genesee group

former turn-around basin for canal boats

Oxford

12

Genesee sandstone cuts

Greene

Sonyea group

oxbow lakes in abandoned meander loops

Tioughnioga R.

Chenango R.

Chenango Forks

Chenango Valley St. Park, kames, kame terraces, kettle lakes, crevasse filling

mostly sandstone and shales

road follows preglacial course of Chenango R., separated from the present course by an "umlaufberg"

Binghamton

Susquehanna R. West Falls group

NY 12
BINGHAMTON—UTICA

The route of the Chenango Canal follows the Chenango River from its junction with the Susquehanna at Binghamton to its headwaters north of Hamilton and then follows north-flowing Oriskany Creek down to the Mohawk Valley. For the most part, the canal bed was laid alongside the river, and only a few sections of river channel were utilized.

The most interesting geological features of the southern Chenango Valley are of glacial origin. In fact, there is no other place in New York that displays such an abundance and variety of features left by stagnant ice.

Binghamton is built on a broad outwash plain of the Susquehanna River at the junction of the Chenango River. Between the junction and Chenango Bridge (5 miles), NY 12 follows the west side of the valley, alongside a delta terrace of sand and gravel that was formed on the sides of the retreating ice tongue.

Between Chenango Bridge and Chenango Forks, the road goes away from the river and behind an umlaufberg, one of those peculiar knobs of rock that are products of ice-marginal drainage changes. The village of Chenango Forks is at the junction of the Tioughnioga River, across the Chenango River from Chenango Valley State Park. The park contains an unusually fine collection of stagnant ice features that are discussed in Interstate 88: Schenectady—Binghamton.

Between Chenango Forks and Greene (8 miles), the river meanders freely over a fairly broad sediment plain largely composed of outwash deposits. Several abandoned meander loops form oxbow lakes. Small cuts near Greene expose gray sandstone of the late Devonian Genesee group. There are kame terraces in several places on the valley sides and a large, triple-crested knob isolated by an ice-marginal channel east of Greene.

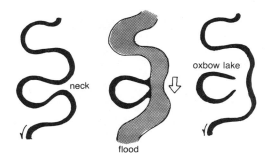

Oxbow lakes form when meander loops are cut off during flooding; when the flood subsides, the river continues along the shortcut route.

Between Greene and Oxford (19 miles), the road is built partly on the old canal bed. This long section of valley contains innumerable mounded deltas and delta terraces and small kettles. Near Brisben is the remarkable Wheeler Brook esker that can be traced for four and a half miles. At its northern end near Brisben it blends into the delta terrace on the side of the valley. From there its winding way can be traced from one erosional remnant to another for nearly three miles downvalley, where it turns south and goes up Wheeler Brook in one continuous ridge a mile and a half long. Eskers are rather rare deposits of rivers that flow under the ice, leaving copious stream-washed sand and gravel. When the ice finally melts, the sediments are dropped in place, and the stream courses are "stenciled" on whatever lies beneath.

During operation of the Chenango Canal, there was a turning basin at Oxford where canal boats tied up. Roadcuts near the village expose brownish gray shales and siltstones of the Unadilla formation of the Genesee group.

One mile north of Oxford, the road crosses a large delta formed in a lake adjacent to the retreating ice tongue by Meadow Creek. Another similar deposit occurs across the river at the mouth of Lyon Brook. Both streams have cut deeply into their deltas since the end of the Ice Age. Ice-marginal terraces are typically clean, water-washed sediments free of clay and are thus ideal for sand and gravel operations. That is why there are so many quarry pits in the Chenango Valley.

Norwich (8 miles north of Oxford) is at the junction of Canasawacta Creek and the Chenango. Between Norwich and North Norwich, the valley floor is wide and flat. Delta terraces are exceptionally well-formed and flat, especially on the east side of the valley. Six miles farther north, at Sherburne, NY 12 leaves the Chenango Valley. The elevation here is 200 feet above the mouth of the river at Binghamton.

Between Sherburne and Waterville (20 miles), the route follows Sangerfield River and skirts the Ninemile Swamp—it's 9 miles long. Waterville lies astride the drainage divide between the south-flowing Sangerfield River and north-flowing Big Creek, which joins Oriskany Creek downstream. The village lies atop the middle Devonian Onondaga limestone which farther west near Syracuse caps the north-facing Onondaga scarp, and farther east blends with the Helderberg scarp. These are just different names for the same Allegheny scarp. Contact with the overlying Hamilton shales is just south of the village.

The road climbs nearly 300 feet higher between Waterville and Paris (6 miles) to the rim of the scarp near the basal contact of the

Onondaga with the Helderberg limestone. From there, the road descends the scarp to Utica (11 miles), dropping more than 1,000 feet to the level of the Mohawk River and crossing band after band of, first, the Helderberg, then the entire Silurian section, and finally the Ordovician rocks of the Mohawk Lowland. Some of the more resistant beds like the Helderberg just north of Paris, and the late Silurian Cobleskill limestone just beneath it, form steps in the face of the scarp.

Schematic cross-section along NY 12 between Sangerfield and Utica

View near Bath of typical glacial trough with flat floor, steep walls, and truncated spurs

VII

THE SOUTHERN TIER EXPRESSWAY

NY 17 crosses the New York section of the Allegheny Plateau province from one end to the other, and in so doing, transects nearly all of the major rock units of the Devonian Catskill Delta. In traveling from east to west, two interrelated trends upwards in the rock section may be seen; these are the rapid decrease in deformation and the change from fine-grained, dark shales and graywackes to coarse-grained, brownish to reddish or greenish sandstones and conglomerates. The steep tilting of the lower Hamilton strata results from the Acadian uplift. The overlying strata are derived from erosion of the Acadian Mountains as they came up, so each younger layer tilts less than the one before. Coarsening of the sediments upwards in the section is related to the growth of the delta throughout the mountain uplift. In its early stages, only the finer erosional debris reached this far westward into a shallow sea. As the building delta displaced the sea and the mountains continued to rise higher in the New England region the shoreline receded westward, and the same depositional site progressed first to seashore and then to alluvial fan on dry land. Coarse grain size and red coloration are both typical of terrestrial deposition, the red resulting from oxidation of iron-bearing minerals to hematite. Brown and yellow-brown colorations reflect the more hydrous environment of the seashore.

The thickest rock section of the Catskill Delta is about 7500 feet in the Catskills, where reddish conglomerates are commonplace. Post-Acadian erosion there, however, has been extensive; the original section is estimated to have been as much as 15,000 feet thick. The uppermost Devonian formations are completely removed, if they were ever present. The delta thins westward to Lake Erie, where it is only about 3000 feet thick. The strata-bound sediments also get finer in that direction, reflecting their increasing distance from the sediment source in the Acadian Mountains.

The Allegheny Plateau of New York experienced innumerable drainage changes influenced by glaciation. Postglacial streams encountered new passageways where divides had been erased, new obstacles in old channels, and changes in gradient. Meltwater drainage was channeled along ice margins in many places, and bedrock valleys were deeply filled with outwash and lake sediments during ice recession. The more obvious regional modifications include a south-pointing system of stream- and glacier-carved valleys in the Finger Lakes West that now drains northward to Lake Ontario; a reversal of drainage in the Allegheny River; and a postglacial Niagara Gorge carved partly in bedrock and partly in a re-excavated pre-Wisconsin gorge that had been buried in till.

Numerous features suggest that glaciation of this section of the Allegheny Plateau contributed greatly to the present condition of the Susquehanna River system. For example, the main river and its largest branches, the Unadilla, Chemung, Cohocton, and Canisteo, together form an arcuate drainage pattern like the rim of a giant wagon wheel with the radiating Finger Lakes as spokes. The glacial history of the Finger Lakes is well documented; and there is strong inference that the modern Susquehanna was established marginal to the same lobate ice sheet that sent tongues through the lake troughs.

At the height of the Wisconsin glacial stage, the ice reached into Pennsylvania, 60 miles south of Binghamton, covering the city site with about 3000 feet of ice. Later, as the glaciers melted back across the plateau, a pause along the line of the river system allowed torrents of meltwater to dig in and

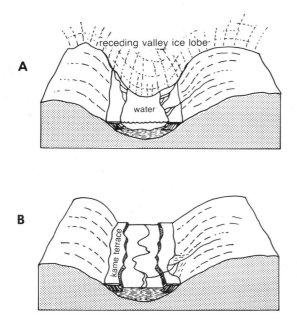

A

receding valley ice lobe

water

B

kame terrace

Anatomy of the underfit Susquehanna River and its tributaries, especially the south-flowing ones. In A, ice marginal streams form a "strip" kame delta along the valley sides as the ice recedes and the river swells with meltwater. B represents the river as it appears today.
Adapted from
P. J. Fleisher, 1977

entrench drainage routes before the Finger Lakes troughs were uncovered. Still later, the receding ice margin stabilized for a long period farther north along the line of the southern tips of the Finger Lakes to form the Valley Heads moraine. All of the major valleys south of the ice front then filled with meltwater that fed into the Susquehanna system. This condition persisted until, or nearly until, the ice front backed off the northern scarp of the plateau, and the Mohawk Valley became the new meltwater escape route. Water levels dropped in the Finger Lakes troughs to below the level of the Valley Heads dams, the lakes were born, the Susquehanna feeder valleys to the south of them virtually dried up, and discharge in the Susquehanna's main branches was drastically reduced.

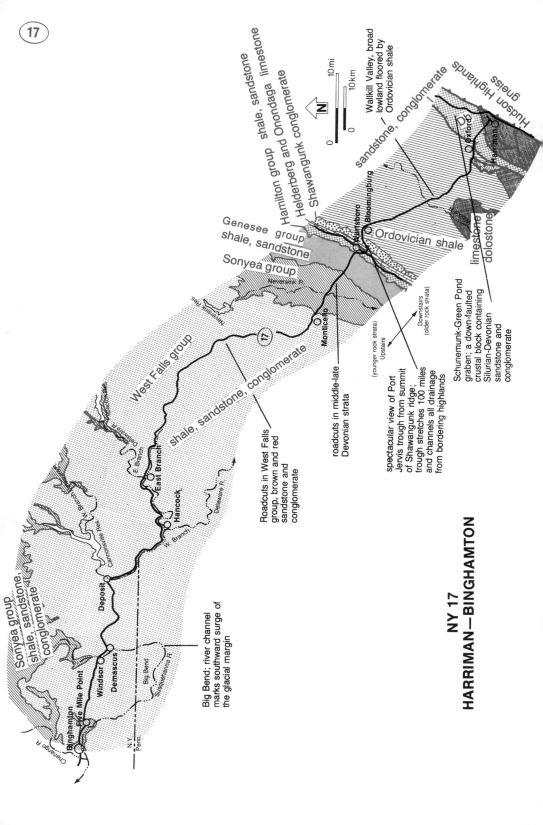

Hamilton group   shale, sandstone   limestone

shale, sandstone

Helderberg and Onondaga

Shawangunk conglomerate

Wallkill Valley, broad
lowland floored by
Ordovician shale

sandstone, conglomerate

Hudson Highlands

gneiss

10 mi

10 km

N

Genesee group
shale, sandstone

Sonyea group

Ordovician shale

limestone
dolostone

Neversink Res.

West Falls group

Neversink R.

Downstairs
(older rock strata)

Upstairs
(younger rock strata)

Schunemunk-Green Pond
graben; a down-faulted
crustal block containing
Silurian-Devonian
sandstone and
conglomerate

Ellenville

Bloomingburg

Wurtsboro

Middletown

Monticello

17

shale, sandstone, conglomerate

spectacular view of Port
Jervis trough from summit
of Shawangunk ridge;
trough stretches 100 miles
and channels all drainage
from bordering highlands

roadcuts in middle-late
Devonian strata

Oxford

Harriman

Sonyea group;
shale, sandstone,
conglomerate

Callicoon Res.

E. Branch

W. Branch

Cannonsville Res.

East Branch

Hancock

W. Branch

Delaware R.

Roadcuts in West Falls
group, brown and red
sandstone and
conglomerate

Deposit

Windsor
Demascus

Five Mile Point

Binghamton

Big Bend

Susquehanna R.

N.Y.
Penn.

Chenango R.

Big Bend; river channel
marks southward surge of
the glacial margin

NY 17
HARRIMAN—BINGHAMTON

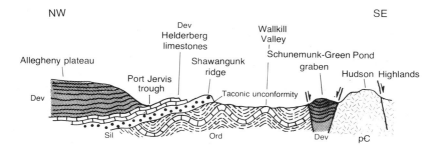

Schematic cross-section along NY 17 from Harriman to Monticello, showing the structural elements mentioned in the text

NY 17:
Harriman—Binghamton
130 mi./210 km.

Harriman is in Cambrian-Ordovician Wappinger group dolostones by the faulted western boundary of the Hudson Highlands. The Highlands are part of the Precambrian "basement" of New York, to which the Fordham gneiss and the Adirondacks also belong. The rocks in this part of the Hudson Highlands are mostly gneisses and quartzites.

Between the Interstate 87 junction near Harriman and Oxford (5 miles), the road passes through the Schunemunk-Green Pond graben, a down-dropped crustal block containing Silurian/Devonian conglomerate and sandstone (also spelled Schunnemunk, Scunnemunk and Skunnemunk). The graben obviously postdates the Taconian block faults of the east side of the Adirondacks and Champlain, Hudson and Mohawk valleys because it cuts rocks formed at a later time. One roadcut near Oxford exposes rusty Hudson Highlands gneiss of a small erosional remnant, or klippe, of a thrust fault. Similar klippen are fairly numerous all along the western border of the Hudson Highlands.

Between Oxford and Bloomingburg (21 miles), you cross the Wallkill Valley, an unusually broad lowland carved by stream and glacier and almost entirely floored by dark Normanskill and Martinsburg group shales, graywackes, and sandstones (see Interstate 84 discussion). Numerous roadcuts reveal folding and pervasive vertical fracturing. The rocks here underwent both Taconian and Acadian deformation.

You cross the Shawangunk Ridge between Bloomingburg and Wurtsboro (4 miles). The ridge rises about 700 feet above the Wallkill Valley and is capped by the Shawangunk formation with steep northwest-dipping strata. The Shawangunk rests on an erosional surface over Martinsburg beds deformed during the Taconian uplift. This is the so-called Taconic unconformity. The tilting of the Shawangunk occurred during the later Acadian uplift, which also overprinted Taconian deformation of the underlying rocks.

The route goes over the Shawangunk Ridge and drops to Wurtsboro at the bottom of the Port Jervis trough, with open views to the Allegheny Plateau to the west. Roadcuts expose steeply dipping Shawangunk sandstone and overlying Helderberg limestone beds. The trough is a major topographic feature of the backside, or dipslope, of the Shawangunk Ridge that stretches for 100 miles from Kingston, New York, to Stroudsburg, Pennsylvania. It has been excavated mainly in weak Hamilton shales and graywackes that rest on the Helderberg limestones (see Interstate 84 discussion).

Between Wurtsboro and Monticello (12 miles), the highway rises about 900 feet onto the eastern scarp of the Allegheny Plateau, passing spectacular roadcuts in middle and late Devonian strata of the Catskill Delta. Dark Hamilton group shales and graywackes near Wurtsboro dip steeply northwestward as in the underlying Helderberg limestones. Upslope are massive gray sandstones and interbedded shales of the Genesee group, and then the massively bedded brownish and reddish sandstones and conglomerates of the Sonyea and West Falls groups. The progression from east to west, therefore, is up the section, from older to younger beds, beginning with the middle Ordovician of the Wallkill Valley. This continues through the Silurian of the Shawangunk Ridge, then units of early Devonian age,

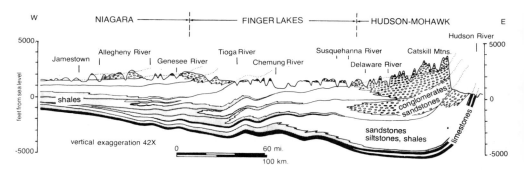

E-W cross-section of Catskill Delta approximately along the line of the Pennsylvania state border.
Adapted from Broughton et al, 1966. Courtesy Kendall / Hunt Publishing Co.

followed by middle Devonian, and finally, late Devonian. This is a rapid journey through the pages of geologic history. The formations continue "younging" farther westward to Lake Erie across the Catskill Delta, but much more slowly because the beds level off. Latest Devonian strata are present only in western New York.

Between Monticello and East Branch (40 miles), there are numerous roadcuts in the West Falls group like those around Monticello: grayish, brownish, and reddish sandstones and conglomerates in flat-lying beds, some with cross-bedding. From the many high points, you will see a streamlined landscape that resulted from glacial scouring. It was first deeply dissected by streams in preglacial time. When glaciers overrode the plateau on four widely-separated occasions during the Ice Age, they deeply gouged the valleys in line with the ice flow, dumped sediments in the cross-valleys, and rounded off the hills. Many of the hills are "till-shadow hills," on which thick drift deposits on south-facing slopes result from glacial deposition on the down-current sides of obstacles, in the same way that water and wind deposits form on the lee side of a rock. Note the prevalence of till wherever the road traverses south-facing slopes and the more frequent bedrock roadcuts on north-facing slopes.

Geologists relate most of the drainage changes of the Allegheny Plateau to the last, or Wisconsin glaciation. Some of the most profound and perplexing changes occurred in the Delaware and Susquehanna river systems of this eastern part of the plateau. This is a classic region of "stream capture," whereby a branch of one drainage system extends itself by erosion of its headwaters until it intersects and captures a branch of another, and thus enlarges its drainage basin. Such "piracy" is most often recognized by barbed drainage junctions that are inconsistent with overall patterns. The west branch of the Delaware River, for example, flows southwestward off the Catskills, parallel to the east branch. The barbed junction is at Deposit, where the river makes a sharp turn to the southeast. Only a few miles southwest of Deposit, across a low divide, is the Big Bend of the Susquehanna, where the river makes a sharp turn from south to northwest toward Binghamton. As a result some Delaware drainage trends are more aligned with the Susquehanna system and vice-versa. Stream capture has undoubtedly taken place here, but the question is which river captured which? The Susquehanna and all of its tributaries have an arcuate pattern that was presumably established along the margin of the Wisconsin ice sheet during a temporary halt in its recession.

The road follows the east branch of the Delaware between East Branch and Hancock (8 miles) where it joins the west branch. It then follows the west branch between Hancock and Deposit (12 miles),

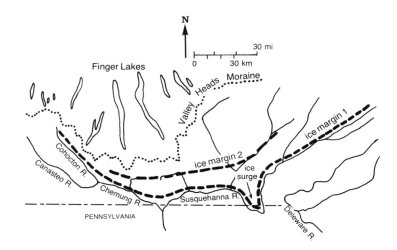

Arcuate trends of the Susquehanna River and its tributaries in relation to probable ice stands during Wisconsin deglaciation, and the similar arcuate pattern of the Valley Heads moraine belt. Adapted from Coates, 1974.

where the barbed junction is located. Between Deposit and Damascus (12 miles), it bridges the divide from the Delaware to the Big Bend section of the Susquehanna, which it then follows for only about 2 miles to Windsor. From Windsor to Fivemile Point (10 miles), you go overland again and rejoin the Susquehanna 2 miles east of Binghamton, on the other side of Big Bend.

All of the roadcuts between East Branch and Fivemile Point are in the West Falls group. Most of the rocks exposed are grayish, thick-bedded to massive sandstones; and many are conspicuously cross-bedded. Reddish and brownish sandstones and conglomerates are less common here than they are farther east. Deep erosion by the Susquehanna and tributaries near Binghamton has exposed the underlying shales of the Sonyea group.

NY 17:
Binghamton—Arkport/Hornell
113 mi./183 km.

All of the principal population centers along this route are situated on wide segments of the Susquehanna Valley or its tributaries, and most are at major stream junctions. These include Binghamton, Johnson City, Endwell, Endicott, Owego, Waverly, Elmira/Horseheads, Corning, Painted Post, Bath, Hornell, and numerous smaller communities. These were attractive sites to the early settlers for a number of reasons, among which perhaps the most important was the use of the rivers as primary transportation routes. Significant also were the abundance of water and the broad, flat, well-drained surface upon which to build and farm.

Most of the major rivers along NY 17 are "underfit," far too small to have carved the large valleys they occupy. Obviously, the amount of water flowing through the valleys must have been greater at one time. The flatness of the valley floor results from deep deposits of glacial outwash sediments; and the broadest segments are by the feeder valleys from the north, many of which now contain only small creeks. The main more or less east-west valleys display numerous ice marginal deposits, such as pitted moraine and outwash, and delta terraces of stream-washed sediments, sometimes called kame terraces.

Several glacial depositional and erosional features are pointed out en route. Roadcuts here, on the other hand, are not terribly exciting to the roadside geologist, consisting almost entirely of flat-lying dark

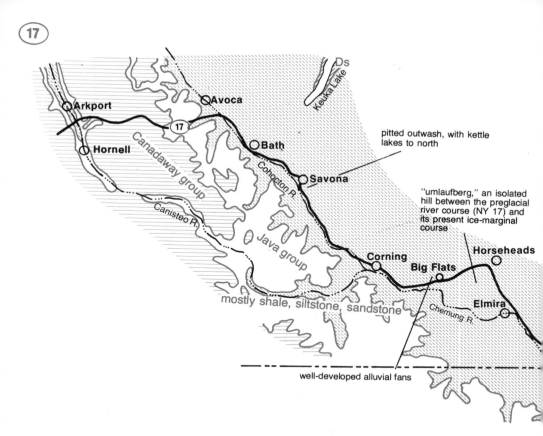

pitted outwash, with kettle
lakes to north

"umlaufberg," an isolated
hill between the preglacial
river course (NY 17) and
its present ice-marginal
course

well-developed alluvial fans

shales, siltstones, and sandstones of late Devonian age. Recognition
of the various Devonian units in the Catskill Delta is mainly by
means of fossils; otherwise, the dark shales of many parts of the
stratigraphic section appear monotonously alike. Fortunately, the
necessary fossils are abundant. Rocks exposed near Binghamton and
Johnson City belong to the Sonyea group, the lowest and oldest
stratigraphic unit in the section between Binghamton and Arkport/
Hornell. Between Johnson City and Bath, all cuts are in West Falls
group, and between Bath and Hornell, the Java and Canadaway
groups are exposed. The trend from east to west, therefore, is upwards
in the stratigraphic section, from older to younger beds. The red and
brown sandstones and conglomerates of the east end of the Catskill
Delta seen between Wurtsboro and Binghamton are almost completely lacking on this section of the Catskill Delta. Sandstone, however, is prevalent in the upper West Falls group toward the west.

Between Binghamton and Waverly (41 miles), the route follows the
Susquehanna downstream with many open views of the valley. The
river meanders widely, but unlike the free-swinging Mississippi,

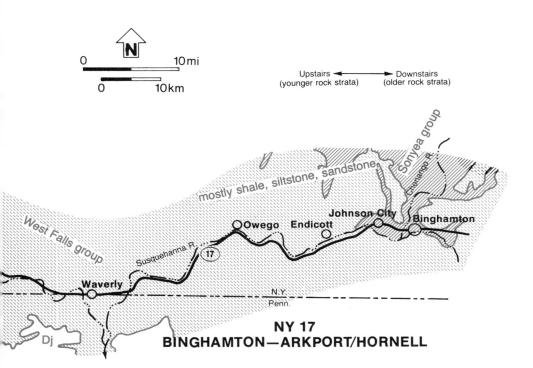

Upstairs ◄────────► Downstairs
(younger rock strata)   (older rock strata)

NY 17
BINGHAMTON—ARKPORT/HORNELL

many of the meanders appear to be in fixed position where they have entrenched themselves over thousands of years. There are innumerable small kettle holes and lakes on the valley floor where stagnant ice blocks were buried in glacial outwash. Delta terraces are visible in many places, especially on the south wall of the valley, but they require a trained eye.

Waverly is at the junction with the Chemung River, which comes in from the west, and where the Susquehanna turns south to cross Pennsylvania to Chesapeake Bay.

Between Waverly and Elmira (18 miles), the highway follows the Chemung River upstream. The river is more underfit than the Susquehanna and meanders more freely between bedrock walls. At about 8 miles west of Waverly, you pass the erosional remnants of a hummocky pitted moraine, best exposed south of the river, which once spanned the valley and formed a dam or divide. It is now completely breached by the river, which flows through at the 800-foot level, or about 150 feet below the crest of the moraine.

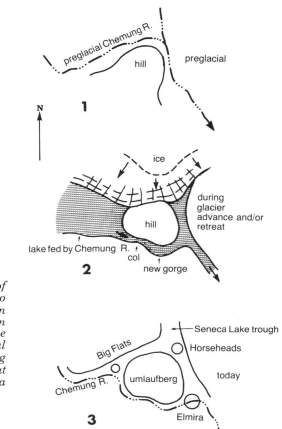

Hypothetical sequence of glacial blockage leading to the formation of an isolated hill called an "umlaufberg" between the pre- and postglacial courses of the Chemung River at Horseheads/Elmira

The contiguous cities of Elmira and Horseheads lie on the broadest and flattest sediment plain of all, where the southern extension of the Seneca Lake trough joins the Chemung Valley. The Valley Heads moraine blocks the trough, beginning only about 3 miles north of Horseheads, and the valley fill at this point is mainly the outwash apron that spreads south of the moraine.

Presently, the Chemung River flows through a narrow bedrock gorge west of Elmira that appears to have been entrenched as an ice-marginal route during Wisconsin deglaciation. The former, pre-Wisconsin, route went farther north via the broad, east-west valley by Horseheads and then south to here. The explanation is that the ice covered the old route for a period when its margin lay along the southern route. Later, when the ice receded farther and uncovered the old route, the river preferred to follow the new course, where it remains even today. The intervening rock barrier is referred to by the German term, "umlaufberg," literally a "run-around mountain." NY 17 follows the old river route through Horseheads.

The route rejoins the Chemung near Big Flats, about 7 miles west of Horseheads, and continues upstream to Corning (13 miles from Horseheads), at the junction with the Cohocton and Tioga rivers. Large, well-developed alluvial fans are on the north wall of the abandoned valley near Big Flats, at the mouths of small side canyons, resembling those of the dry, southwestern United States. Such fans occur at many of the small side canyons elsewhere, but here they have not been destroyed by river erosion.

Between Corning and Bath (20 miles), you continue along the Cohocton Valley. The main valleys between Horseheads and Bath are exceptionally broad and trough-like; and the rivers are more underfit than those farther east. Steep valley sides end abruptly against the flat, sediment-filled floors. North of Corning, several small tributary valleys visible from the highway have the distinctive U-shape produced by glacial erosion. These are really "hanging valleys," hollowed out by ice tongues tributary to the main one, with bedrock floors high above that of the buried main valley. The village of Savona between Corning and Bath lies on a pitted outwash plain 3 miles south of a fine group of large kettle lakes in Mud Creek Valley. Bath is in a similar location, at the junction of the Cohocton Valley with the southern extension of the Keuka Lake trough.

Between Bath and Arkport/Hornell (22 miles), NY 17 leaves the Cohocton and goes overland to the Canisteo Valley, affording open views of the upland topography of this section of the plateau. It is not flat, as you might expect for a plateau, but rolling, with considerable relief. Angular forms are absent and the landscape has been streamlined by the scouring action of overriding ice. It is a lovely, peaceful sort of terrane.

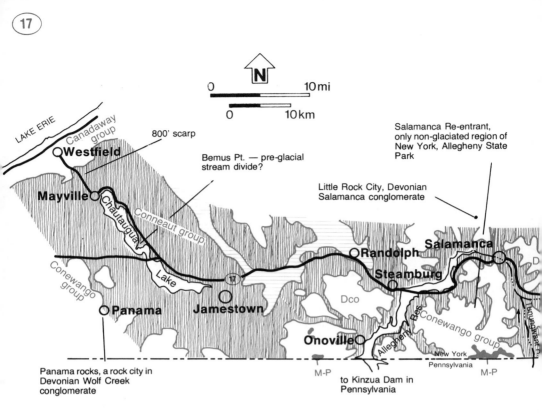

N

0   10 mi

0   10 km

LAKE ERIE

Canadaway group

800' scarp

Bemus Pt. — pre-glacial stream divide?

Salamanca Re-entrant, only non-glaciated region of New York, Allegheny State Park

Little Rock City, Devonian Salamanca conglomerate

Westfield

Mayville

Chautauqua

Conneaut group

Conewango group

Lake

Salamanca

Randolph

Steamburg

D

17

Dco

Panama

Jamestown

Onoville

Allegheny Res.

Conewango group

New York

Pennsylvania

Tunungwant Cr.

Panama rocks, a rock city in Devonian Wolf Creek conglomerate

M-P

to Kinzua Dam in Pennsylvania

M-P

**NY 17
ARKPORT/HORNELL—WESTFIELD**

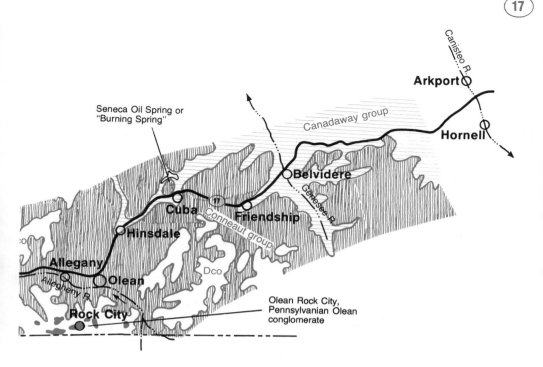

NY 17:
Arkport/Hornell—Westfield
128 mi./207 km.

Between Arkport/Hornell and Belvidere (24 miles), NY 17 crosses the stream-dissected divide between the Susquehanna and Genesee River basins. The route follows several small, fairly deep valleys that obviously lack the broad, flat floors and abrupt, steep walls of glacial troughs and sluiceways. Bedrock exposed in very few cuts is flat-lying, late Devonian shales and siltstones of the Canadaway and Conneaut groups.

Belvidere is in a broad stretch of the Genesee Valley. The Genesee is the only major river that completely crosses the state. It is also the only one that maintains its pre-Wisconsin northward flow despite diversions and obstructions introduced by glaciation. The present course lies partly in re-excavated, preglacial valleys and partly in those newly carved in bedrock. The river heads up in numerous small streams just over the border in the hills of Pennsylvania. These coalesce to a full-fledged river at the appropriately-named village of Genesee, Pennsylvania, near the state line. Belvidere is in the middle of the 60-mile segment of the river between Genesee and Portageville, where the valley is broad and flat, the gradient is gentle, and the river meanders lazily back and forth. Between Portageville and Mt. Morris, the scene is dramatically different in the deep bedrock gorges and waterfalls of Letchworth State Park (see Letchworth State Park discussion). Between Mt. Morris and Rochester, the river assumes its laziest character, meandering even more widely than before. Finally, passing through the Rochester Gorge to Lake Ontario, it drops over three more waterfalls and transects Silurian and Ordovician bedrock similar to that of the Niagara Gorge.

Between Belvidere and Cuba (12 miles), you cross overland to Friendship, and then follow Van Compen Creek and Tannery Creek valleys between Friendship and Cuba. Cuba is on the wide, flat floor of the Oil Creek/Black Creek Valley, which obviously served as one of the main drainage routes during Wisconsin deglaciation (see NY 17: Binghamton—Arkport/Hornell discussion). This area is most famous for the Seneca Oil Spring, the "burning spring" where oil was first discovered in America in 1627, more than 230 years before the world's first successful oil well was drilled at Titusville, Pennsylvania. The spring is now protected in a small park in the Oil Spring Reservation two miles northwest of Cuba.

The Oil Creek/Black Creek Valley is also the site of the abandoned Genesee Valley Canal, perhaps the most intriguing of the many canals built in New York in the mid-1800s. This one is called a "lateral canal" because it cuts across the geologic structures that, more than anything else, controlled canal routing. Completed in 1861, this canal provided a link between Lake Ontario at Rochester and the Allegheny River at the Pennsylvania border. One hundred and twelve locks were required in 124 miles. The Letchworth Gorge

Seneca oil spring

section was the most difficult engineering accomplishment, with 50 locks in the short distance between Sonyea and Portageville, and parts of the canal notched into the walls of the gorge above the river! Commercially, it was one of the least successful of the New York canals for a number of reasons. It was, first of all, expensive to operate and maintain because it had so many locks. Parts of it in Letchworth crossed unstable landslide areas where repairs were often required. There were a lot of other reasons, but the early demise of the Genesee Valley Canal and many others was mostly due to competition from the railroads, which began providing a cheaper and faster means of transportation soon after they were built. The railroads have since added insult to injury by laying new lines along parts of the abandoned canal beds.

Southwest of Cuba, you follow Oil Creek to its junction with Ischua Creek at Hinsdale (7 miles). Between Hinsdale and Olean (5 miles) you follow Olean Creek to its junction with the Allegheny River. Each of the creeks is underfit, meandering widely on a broad, flat floor between steep bedrock walls. These are typical sediment-filled glacial troughs.

The Allegheny River enters New York State from Pennsylvania southeast of Olean, flows by Olean to Salamanca, then circles around the west side of Allegany State Park before re-entering Pennsylvania south of Onoville. This roughly triangular area enclosed by the river and the state line, as well as its Pennsylvania extension, is known geologically as the Salamanca re-entrant. It is the only area of New York that escaped glaciation during the Ice Age, and furthermore, it is the most northerly region of unglaciated landscape in eastern North America. As a result, the erosional surface of the re-entrant differs markedly from that to the north. Most striking is the total absence of glacial troughs and their associated features, such as glens, hanging valleys, finger lakes, etc. In their place is a landscape rather evenly and deeply dissected by an intricate tree-like network of small streams and a more angular topography that has not been subdued and streamlined by overriding ice.

Glaciation drastically changed the Allegheny River system. The present river drains southward to the Ohio River and then to the Mississippi. The preglacial river flowed northward to the St. Lawrence via the Erie and Ontario basins. One of the keys to the alteration was the erosional breaching of the Kinzua col, a preglacial drainage divide located at the present site of the Kinzua Dam in the Big Bend section of the river in Pennsylvania. The Gowanda and Steamburg moraines, formed during Wisconsin deglaciation, also prevented the postglacial river from finding its old routes to Lake Erie. The major changes are shown on the accompanying preglacial

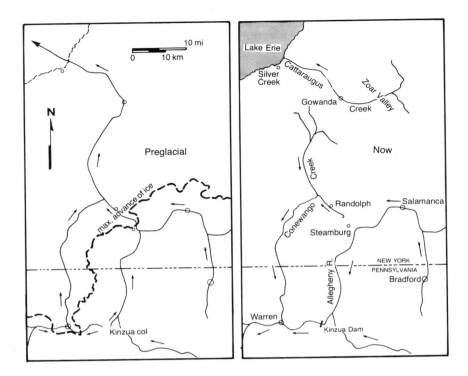

Changes in the Allegheny River system induced by glaciation. The Kinzua Col, a drainage divide, was worn away by the swollen river during deglaciation. Moraine at Steamburg and Gowanda also blocked the old stream course. Headwater erosion by the postglacial Cattaraugus Creek resulted in the deep gorge of Zoar Valley. Courtesy Kendall/Hunt Publishing Co.

and postglacial drainage maps.

Between Olean and Steamburg, (29 miles), the route follows the broad Allegheny Valley downstream. En route, you pass three glacial troughs on the north: Fivemile Creek valley by Allegheny, Great Valley by Salamanca, and Little Valley just west of Salamanca. The Allegheny Valley itself is deeply buried in outwash sediments released from the ice as it melted. It differs from the glacial troughs in that it has never been in direct contact with the ice. At maximum, the Wisconsin ice margin lay a few miles outside the valley perimeter.

The Allegany State Park region is unique for other reasons, too. It is the only part of New York where strata of Mississippian and Pennsylvanian age are found. It is also the center for oil production in the state. There are many hundreds of wells shown as small open circles on the 7.5 minute topographic quadrangle maps of the United States Geological Survey. Most of these are inactive, having been pumped dry years ago, but production from active wells as late as 1973 contributed well over $5 million to the state's economy. The first oil well in New York was drilled in 1865 near the village of Limestone at the southeastern corner of Allegany Park. The oil comes almost exclusively from permeable sandstone lenses in the late Devonian Conneaut and Canadaway groups. There is natural gas in this region as well, and some old gas wells have been used for storage of gas imported from other states.

There are a number of conglomerate beds in this region, in the late Devonian Conewango group, the Mississippian Pocono group, and the Pennsylvanian Pottsville group, that are of special interest because they form rock cities. Where these massive beds crop out in hillslopes, as most of them do, they break up into huge, joint-bounded blocks that tumble and lean against each other so as to produce narrow passageways, streets, and caves. The best known and most accessible is Olean Rock City, south of Olean, in the Pennsylvanian Olean conglomerate. Others are Little Rock City, north of Salamanca, Bear Caves, within the park, in the Devonian Salamanca conglomerate, and Thunder Rocks, within the park, in the Pennsylvanian Olean conglomerate. Farther afield, Panama Rocks, near the village of Panama and west of Jamestown, are formed from the Wolf Creek conglomerate member at the base of the Conewango group.

Most of the way between Steamburg and Jamestown (21 miles), you follow the exceedingly broad Conewango Valley, the former main branche of the preglacial Allegheny. The east branch heads up on the Steamburg, or Kent, moraine, and flows northwestward to beyond Randolph, where it joins the west branch from the northwest. The main branch then continues toward Jamestown, then south to War-

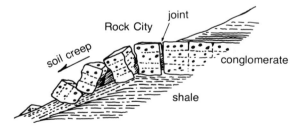

This is how "rock cities" form. Massive, resistant conglomerate beds break into huge, joint-bounded blocks as erosion undermines their weak shale foundation, and are slowly carried downslope by soil creep. The "buildings" lean this way and that, with narrow "streets" and tunnels between them.

*Rock City Park near Olean;
massively bedded
conglomerate*
Courtesy Kendall / Hunt Publishing Co.

ren, Pennsylvania, where it joins the Allegheny. Flow direction in the west and main branches is now opposite to the preglacial flow.

Jamestown is at the east end of Chautauqua Lake, a finger lake of a different sort. The lake stretches for 25 miles in a northwesterly direction and averages about 2 miles wide. In plan view, it almost looks like two lakes because there is an unusual constriction in the middle at Bemus Point, where the new (1982) bridge carries NY 17 across the lake. The constriction is probably the site of a preglacial drainage divide, from which one river flowed northwestward to the Erie basin, and the other southeastward to the ancestral Allegheny. Glaciation scooped out the valleys, wore down the divide, and later dammed up the northwest end with moraine. The lake drains at the east end via the Chadakoin River through Jamestown to Cassadaga Creek, and then to Conewango Creek. Cassadaga Creek occupies another extremely broad, flat, glacial trough that trends northwest parallel to Chautauqua Lake.

Between Bemus Point and Mayville (10 miles), you follow old NY 17 high on the north side of the lake with many splendid views. The lake was a favorite route for French explorers, shortening the portages between Lake Erie and the Ohio River. The upland here is more gently glaciated than that of the Finger Lakes region to the east.

Between Mayville and Westfield (7 miles), old NY 17 descends more than 800 feet over the northwest scarp of the Allegheny Plateau, with some spectacular panoramic views of the Erie basin. The road near Westfield is perched on the narrow divide between the deep bedrock gorges, or "gulfs," of Chautauqua Creek and Little Chautauqua Creek. Exposed in the gorge walls are dark shales and siltstones of the late Devonian Canadaway group.

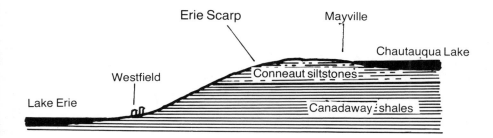

Cross-section along old NY 17 between Mayville and Westfield, showing Erie scarp

175

Indian Ladder Trail on the Helderberg scarp at John Boyd Thatcher State Park, showing cliff of Helderberg limestones. Courtesy Kendall/Hunt Publishing Co

VIII

THE CENTRAL CORRIDOR

The human history of the Mohawk Valley is long and varied, largely because it is such a natural passageway and really the only easy route from east to west, a reflection of the geology. The most ancient human implements discovered there are some Clovis fluted points which, from dating of similar finds in the southwestern United States, may be as old as 10,000 years. More recently the territory was the home of the Mohawks, the easternmost segment of the Iroquois Nation. The Mohawks were a savage and cannibalistic tribe which resented any intrusion by the white man, or even by their Iroquois cousins. Early white trespassers, mostly missionaries, trappers, and explorers, were often tortured, killed, and eaten.

In the late 17th and early 18th centuries, the eastern part of the valley was settled by Dutch and German colonists from the Hudson Valley, who managed to live among the Mohawks and even intermarry with them. Slow colonization continued through much of the 18th century, in favor of the British rather than the French, for whom the Mohawks had an intense hatred.

During the American Revolution, control of the valley by the British was seen as essential to quell the rebellion. At this they failed. The decisive battle was fought at Oriskany at the western end of the valley in 1777, where the American patriots, under the command of General Nicholas Herkimer, defeated the combined British, Tory, and Mohawk forces as they attempted to advance eastward through the valley. This, combined with the defeat of General Burgoyne's army in the Battle of Saratoga in that same year, eventually led to surrender of the British at Yorktown, Virginia, in 1781. A new nation was born.

The Mohawk Valley was the obvious choice for routing of the eastern segment of the original Erie Canal: it has a broad, even-graded, easily-excavated floodplain; and over most of its length, the river itself was large enough to accommodate barge traffic with minimal dredging and modification. At the western end, where the river is small, the canal was dug alongside it, where you see it today.

This is also the obvious route for Interstate 90, which follows the valley all the way to Schenectady. From its position on the valley sides or floor, you have ample opportunities to see why the Mohawk region is considered one of the state's greatest scenic pleasures.

The dominant topographic character of this section of New York, as it is throughout the southern tier, is the tilted mesa. Paleozoic strata of Ordovician, Silurian, and Devonian age dip gently to the south, and long-term erosion has produced east-west trending, north-facing scarps on the upturned ends of the more resistant beds. Near Buffalo, there are three principal scarps, arranged in steps that descend northward from the Allegheny Plateau in order: the Portage, Onondaga, and Niagara. The broad treads of these steps in western New York are relatively flat erosional surfaces. The highest, the Erie plain, lies between the Portage and Onondaga scarps, broadens westward to include the shallow Lake Erie depression, and is underlain by rocks of Devonian age. The middle, or Tonawanda plain, lies between the Onondaga and Niagara scarps, and is floored by Silurian rocks. The Ontario plain lies north of the Niagara scarp, includes Lake Ontario, and is underlain in New York by Ordovician rocks. The Niagara scarp, a 250-foot high cliff/slope north of Buffalo, gradually flattens eastward and is imperceptible beyond Rochester. The Erie plain narrows eastward and disappears near Syracuse as the Onondaga and Portage scarps merge. Therefore, the three-fold division of plains

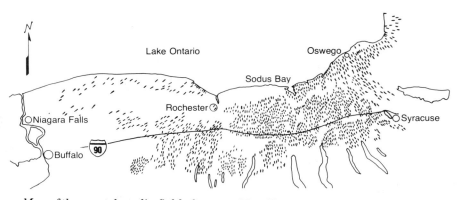

Map of the great drumlin field of western New York, and the Interstate 90 route through it. The drumlins radiate from the Ontario shore because the lake basin served as a spreading center for Wisconsin ice.
Courtesy Kendall/Hunt Publishing Co.

becomes just two-fold near Rochester, and only one, the Ontario plain, exists near Syracuse north of the Onondaga scarp.

West of Syracuse, Interstate 90 and US 20 take you through the heart of one of the greatest drumlin fields in the world. In the region roughly confined by Lake Ontario and the Finger Lakes, and stretching from Syracuse westward for about 100 miles, more than 10,000 of these peculiar glacial features have been counted. Drumlins are streamlined hills of glacial drift that have been molded by overriding ice. They are always elongated in the direction of ice movement and tend to have an ideal long profile rather like an airplane wing with its blunt end pointed upglacier, and long trailing slope downglacier. As you can see on the glacial map in the front of the book, the linear forms radiate from Lake Ontario, indicating that the lake basin served as a spreading center for the ice. The field continues northward toward Watertown around the east end of the lake. Thousands of drumlins on the Canadian side of the lake also belong to this enormous field.

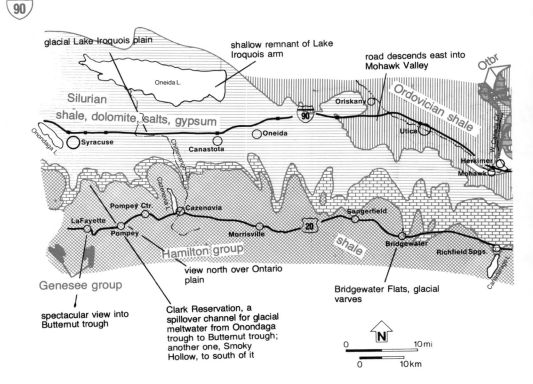

glacial Lake Iroquois plain

shallow remnant of Lake Iroquois arm

road descends east into Mohawk Valley

Otbr

Oneida L.

Ordovician shale

Silurian
shale, dolomite, salts, gypsum

Oriskany

Oneida

Utica

Onondaga L.

Syracuse

Canastota

Chittenango L.

90

Herkimer

Mohawk

W. Canada Cr.

Cazenovia L.

Pompey Ctr.

Cazenovia

Sangerfield

LaFayette

Pompey

Morrisville

20

Bridgewater

shale

Richfield Spgs.

Hamilton group

Genesee group

view north over Ontario plain

Bridgewater Flats, glacial varves

Canadarago L.

spectacular view into Butternut trough

Clark Reservation, a spillover channel for glacial meltwater from Onondaga trough to Butternut trough; another one, Smoky Hollow, to south of it

N

0   10 mi

0   10 km

Interstate 90
SYRACUSE—MASSACHUSETTS STATE LINE

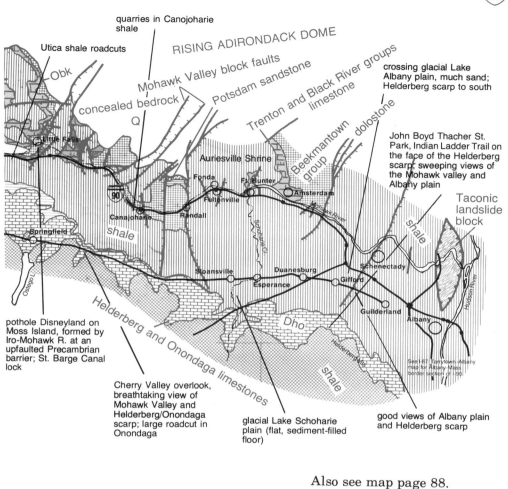

quarries in Canojoharie shale

Utica shale roadcuts

RISING ADIRONDACK DOME

Mohawk Valley block faults

Potsdam sandstone

Trenton and Black River groups

limestone

Obk

concealed bedrock

Q

Beekmantown group

dolostone

crossing glacial Lake Albany plain, much sand; Helderberg scarp to south

John Boyd Thacher St. Park, Indian Ladder Trail on the face of the Helderberg scarp; sweeping views of the Mohawk valley and Albany plain

Little Falls

Auriesville Shrine

Fonda

Ft. Hunter

Amsterdam

Mohawk River

Taconic landslide block

90

Fultonville

Canojoharie

Randall

shale

Springfield

Schoharie Cr.

shale

Otsego L.

Sloansville

Duanesburg

Esperance

Gifford

Schenectady

Hudson River

Helderberg and Onondaga limestones

Dho

Guilderland

Albany

Helderberg scarp

shale

pothole Disneyland on Moss Island, formed by Iro-Mohawk R. at an upfaulted Precambrian barrier; St. Barge Canal lock

See I-87 Tarrytown-Albany map for Albany-Mass border section of I-90.

Cherry Valley overlook, breathtaking view of Mohawk Valley and Helderberg/Onondaga scarp; large roadcut in Onondaga

glacial Lake Schoharie plain (flat, sediment-filled floor)

good views of Albany plain and Helderberg scarp

Also see map page 88.

Interstate 90:
Syracuse—Massachusetts State Line
177 mi./287 km.

THROUGH THE HISTORIC MOHAWK VALLEY

Between Syracuse and Oriskany, the landscape is quite flat, in contrast to that of the great drumlin field that begins a few miles west of the city. This is lake plain developed during the late stages of Wisconsin glaciation, when glacial Lake Iroquois extended a long arm eastward to drain into what was then the head of the Mohawk

Valley at Little Falls. The St. Lawrence River then was still blocked by ice, so that the Mohawk served as the main outlet for all of the newly-forming Great Lakes, whose waters passed through Lake Iroquois. Oneida Lake, which lies a few miles north of the highway between Syracuse and Canastota (21 miles), is a shallow remnant, and all that remains of the arm of Lake Iroquois. The lake now serves as part of the State Barge Canal, and drains at its western end to the Oswego River, and then to Lake Ontario at Oswego.

The Onondaga-Helderberg scarp, meanwhile, is clearly visible a short distance to the south. Near Oneida, it displays pronounced tilted mesa steps developed on the more resistant layers (see Interstate 81: Syracuse-Alexandria Bay discussion of north end of Tug Hill).

Near Oriskany, the road descends into the Mohawk Valley and crosses the river and the State Barge Canal. The Mohawk heads up at Delta Reservoir, a few miles north of Rome, located where the valley blends into the Lake Oneida (Iroquois) plain. Throughout its 100-mile length, the river is but a shadow of its great predecessor, called the Iro-Mohawk in reference to its origin at glacial Lake Iroquois. Here it is little more than a trickle that meanders back and forth across a broad flood plain. It is said to be "underfit," meaning that it flows through a valley much too large to have been carved by such a small stream. The immensity of the Iro-Mohawk Valley can best be appreciated a little farther downstream, east of Utica, where Dutch Hill abruptly rises 600 feet from the flood plain to form the south wall of the valley. Over much of its length, the south wall is notably steeper than the north wall because the strata dip, or slope, slightly southward. At Herkimer, the south wall is especially steep and high where the valley skirts the base of the Onondaga-Helderberg scarp (see NY 28: Oneonta—Boonville discussion).

Immediately east of Herkimer, the road climbs 700 feet up the south side of the valley and then back down to the river beyond Little Falls at Exit 29A, passing several large roadcuts in dark Ordovician Utica shale, some showing small folds and faults. The exposure at mile 212.8 is cut by the Little Falls fault, one of the numerous large up-down faults that transect the Mohawk Valley. Where the fault crosses the river 2 miles to the northeast, the eastern side has dropped about 1000 feet, placing the weak Utica shale against hard Precambrian gneisses on the west. This is one of the most significant features in the erosional evolution of the modern Mohawk Valley.

In preglacial times, the gneisses and immediately overlying Ordovician Little Falls dolostone formed a topographic barrier, a drainage divide from which the ancestral Mohawk flowed eastward, and the so-called Rome River drained westward. The barrier was worn

Pothole "Disneyland" at Moss Island, a product of torrential flow and deep erosion of an upfaulted barrier of Precambrian rock by the Iro-Mohawk River
Courtesy Kendall / Hunt Publishing Co.

away when Lake Iroquois stretched its long arm to here, spilling a torrent of water into the Mohawk Valley and forming the Iro-Mohawk River. By the time the alternate St. Lawrence River escape route was finally freed of ice, a deep slot had been cut through the Precambrian gneisses, and the west to east drainage direction was well-established. The walls of the gorge now rise about 520 feet on the south and 700 feet on the north. One of the feature attractions is the "Disneyland" of huge potholes on the eastern end of Moss Island, located in Little Falls between the river and Lock 17 of the canal. This incredible profusion of potholes, now recognized as a National Historic Landmark by the National Park Service, is the most tangible evidence of the enormous river flow that once passed this way.

The eastern edge of the Precambrian block at Little Falls is marked by a steep fault scarp, and an abrupt widening of the river floodplain at German Flats just beyond the jaws of the gorge. Near river level, several erosional remnants of flat-topped river terraces may be seen on either side of the valley, elevated 100 feet or more above the present floodplain. A lot of sediment was dumped here as the Iro-Mohawk disgorged into this broad, relatively quiet stretch of water that was later to be dissected by downcutting of the modern Mohawk.

Between Little Falls and Canajoharie (15 miles), the highway passes over several more transecting faults, but none with the same striking topographic expression. Roadcuts expose Ordovician limestones and dolostones of the Trenton, Black River, and Beekmantown groups, as well as Utica shale. East of Canajoharie, you have excellent open views of the valley. The bridge-like structure over the river at Canajoharie is actually an "ice dam" used to maintain a 14-foot water depth during the canal season, but removed during the winter or spring to prevent damage from ice jams. This is one of 10 in the canal system.

The Little Falls' features are repeated at Randall, 7 miles east of Canajoharie, where the Noses fault crosses the river. Again there is a narrow, deep cut through the upthrown fault block, and a sudden valley widening where the gorge ends at a north-south fault scarp. "Big Nose" rises steeply 560 feet above the north side of the river in the center of the cut, while "Little Nose" rises over 400 feet on the south side only a half mile away. Precambrian gneisses are exposed along the railroad tracks on the south side of the gorge between miles 189-188, with late Cambrian Little Falls dolostone resting on top of them. The contact, like that of the Alexandria Bay roadcut (see NY 38, 12, 12E: Ogdensburg—Cape Vincent—Watertown discussion), apparently represents an unconformity with a 600 million-year gap in the rock record.

The wide flat-topped plain east of the fault scarp is called the Fonda Wash plain. This is an Ice Age outwash delta and lake plain that stretches 3.5 miles northeast to Fonda, and 4.5 miles from the river northward. The plain is dotted with several sand-gravel quarries.

Between Fultonville-Fonda and Fort Hunter, you pass the Auriesville Shrine, dedicated to some of the martyred white men who died at the hands of the Mohawks. Schoharie Creek, the only large stream that crosses the Helderberg scarp from south to north, enters the Mohawk River at Fort Hunter (see US 20: Lafayette—Guilderland discussion). The pronounced terracing visible at the Shrine results from the meandering and downcutting of the modern stream into thick lake and delta sediments left during the Lake Albany stage of Wisconsin glacial retreat, actually Lake Amsterdam in this part of the Mohawk Valley.

At the top of the rise out of the valley, the Helderberg scarp is visible to the south as it is at other high points farther east. At Exit 27 to Amsterdam, there are deep cuts through Canajoharie shale into

Black River limestone. At Pattersonville, the highway crosses another one of the major up-down faults that transect the valley, this one referred to as the Hoffmans fault. The east-facing scarp here is not as well-defined nor as high as those of the Little Falls or Noses faults; however, the channel fill east of the scarp is known to be 200 feet deep, and the scarp is believed to have been the site of a 100-foot waterfall in preglacial time.

Between Exits 27 and 26, the road trends southeastward away from the river and closer to the Helderberg Mountains. Between Exits 26 and 25, there are several large cuts in Ordovician shales, near John Boyd Thacher State Park, which lies atop the scarp. Farther east is the great Schenectady delta which was formed by the Mohawk River during the Lake Albany stage, and is evident in the many sandy roadcuts. Much of the delta sediment derives from the Fonda Wash plain and Lake Amsterdam deposits that were reworked and transported to this point after the mouth of the Mohawk was free of ice.

Traveling over the Lake Albany plain, Interstate 90 passes around the north side of Albany, crosses the Hudson River, and then goes south for 12 miles, giving good views of the Hudson Valley. Turning east again, the route continues for 17 miles to the Massachusetts border across the Taconic Mountains, with geology dramatically different from that of the Mohawk Valley. The so-called "Taconic Klippe" is an enormous mass of Cambrian and Ordovician rocks, which either slid in titanic landslides or was shoved westward along great thrust faults from the New England region during the Taconian mountain-building event, and then isolated by erosion. The rocks are intensely faulted, folded, and metamorphosed with increasing grade of metamorphism to the east. Most of the roadcuts you see along the way consist of crumpled dark-colored slates and phyllites. Near the

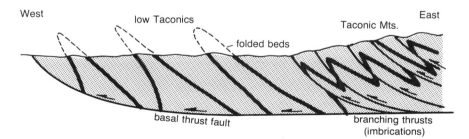

Schematic east-west cross-section through the Taconic Klippe near Glens Falls showing tight folding of beds and stacking of branching thrust sheets over a basal thrust, like "folded bridge chairs" against a wall

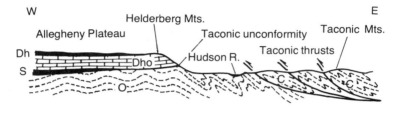

toll booth near the border, there is an excellent exposure of an anti-clinal fold with uparched layers on the south side of the highway. Highly deformed light gray marbles are also exposed in cuts near the exit. The topography, as usual, reflects the bedrock geology: the ridges trend north-south parallel to the many faults and folds; the hills are made up of the more resistant slates and phyllites; and many valleys are underlain with weaker marble. The landscape also has a distinctive hummocky character that reveals the crumpling of the bedrock layers.

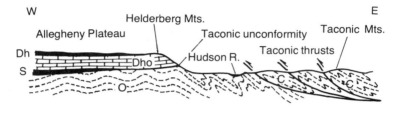

Schematic E-W cross-section a few miles south of Albany

Drumlin near Syracuse   Courtesy Kendall/Hunt Publishing Co.

Interstate 90:
Syracuse—Buffalo
135 mi./219 km.

THE GREAT DRUMLIN FIELD
OF WESTERN NEW YORK

This is one region of New York where topography is dominated by glacial forms rather than bedrock; another is Long Island. The streamlining action of glaciation produced a smooth rollercoaster surface. This is rich farmland, and the inherent tranquil beauty of the landscape is enhanced by the even furrows of plow and crop.

Between Exits 36 and 39, the highway skirts the north shore of Onondaga Lake and crosses the Barge Canal outlet. This branch of the canal connects with the main Seneca River branch, which, downstream, joins the Oneida River to Oneida Lake, and the Oswego River branch to Lake Ontario. Onondaga Lake differs from the main

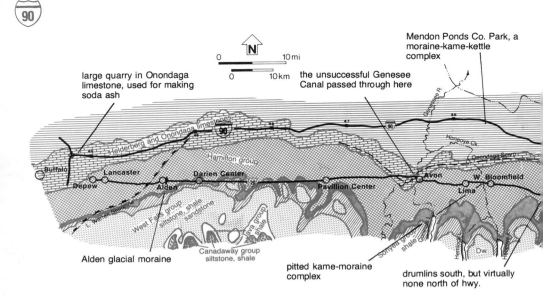

large quarry in Onondaga limestone, used for making soda ash

the unsuccessful Genesee Canal passed through here

Mendon Ponds Co. Park, a moraine-kame-kettle complex

Buffalo
Lancaster
Depew
Alden
Darien Center
Pavillion Center
Avon
W. Bloomfield
Lima

Helderberg and Onondaga limestones
Hamilton group
West Falls group siltstone, shale, sandstone
Java group shale
Canadaway group siltstone, shale
Sonyea group shale
Onondaga Scarp
Honeoye Ck.

Alden glacial moraine

pitted kame-moraine complex

drumlins south, but virtually none north of hwy.

Interstate 90
SYRACUSE—BUFFALO

Finger Lakes in that it lies entirely within the Ontario Lowlands province rather than in the Allegheny Plateau (see Interstate 81: Syracuse—Pennsylvania Border discussion). The drumlins begin at about mile 290 and continue in great profusion to the Montezuma Marshes at mile 313. Their long axes trend generally north-south so that their nearly parabolic cross-sections can be seen in roadcuts, while longitudinal profiles may be viewed ahead or behind. The route follows a lowland strip underlain by weak Silurian shales, salt, and gypsum just north of the abandoned original Erie Canal route and south of the more meandering Seneca River route of the Barge Canal. The bedrock does influence the topography, despite the cover of glacial debris. The Onondaga scarp forms the drainage divide a few miles to the south, whereas the very low ridge of resistant Silurian Lockport dolostone forms the divide to the north. The Lockport becomes much more prominent westward where it caps Niagara Falls and the Niagara scarp. Here the glacial scour appears to have been much more severe than in the west, resulting in greater erosional reduction of the east-west-trending divides.

The Montezuma Marshes lie in a northward extension of the Cayuga Lake trough (see US 20: Canandaigua—LaFayette discussion). This extensive region of ten square miles of bogs and ponds is

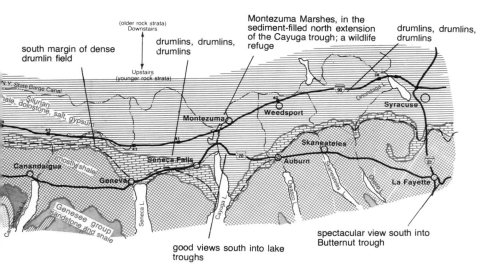

(older rock strata)
Downstairs

Montezuma Marshes, in the
sediment-filled north extension
of the Cayuga trough; a wildlife
refuge

drumlins, drumlins,
drumlins

south margin of dense
drumlin field

drumlins, drumlins,
drumlins

Upstairs
(younger rock strata)

N.Y. State Barge Canal

shale, dolostone, salt, gypsum
Silurian

Syracuse

Onondaga L.

Montezuma

Weedsport

mostly shale

Canandaigua

Seneca Falls

Skaneateles

Geneva

Auburn

La Fayette

Genesee group
sandstone and shale

Seneca L.

Cayuga L.

Owasco

Skaneateles

Otisco

spectacular view south into
Butternut trough

good views south into lake
troughs

now under Federal protection as the Montezuma National Wildlife
Refuge.

More drumlins occur west of the marshes. Most of the roadcuts
through the drumlins are grassed over so that the nature of their
materials is not readily apparent; stopping to examine them at closer
range is not permitted on the Thruway under any circumstances.
Most are made up of heterogeneous, unstratified, glacial till, sometimes with a wide range of particle sizes, from clay to large boulders, a
mixture that is stable in steep slopes. Most of the particles have been
plucked by the glacier from bedrock of the Ontario Lowlands, but
"erratic" boulders of Precambrian rock from Canada may also be
present.

Molding of this stiff conglomeration of glacial debris into drumlins
occurs under moving ice, and it has been shown experimentally that
very special conditions of pressure, hence thickness of overriding ice
and glacier velocity, are required. (See Ice Age chapter.) As a result,
drumlins are unusual; and their presence here in such incredible
density and perfection of form is truly rare. A good place to see the
materials and internal structure of the drumlins is along the Lake
Ontario shore between Sodus Bay and Oswego, where water and
badlands erosion have cut into them.

West of Exit 42, the highway skirts the Onondaga scarp and then
follows the valley of the Canandaigua Lake outlet and the sharply
defined southern boundary of the drumlin field. There are almost no
drumlins immediately to the south, where the land rises gently to the

189

scarp; this demonstrates the remarkable sensitivity of the drumlins to their physical environment. Where the ice moved against the slope, the pressure at its base was apparently too great, and the existing till was either bulldozed away or smeared.

Between Exits 45 and 46, you will pass north of Mendon Ponds County Park. This is a complex of glacial features that marks a stand of the Wisconsin glacier front about 11,000 years ago, younger than the Valley Heads moraine farther south, and older than a similar complex at Pinnacle Hills at the south side of Rochester. In addition to moraines, other types of deposits are especially well-displayed here. Kames are mounded outwash sediments dumped next to the ice. Kettles are hollows where stray ice blocks were partly or wholly buried in outwash before melting. Eskers are long sinuous ridges of sand and gravel that mark the courses of rivers under or on the ice.

West of Exit 47, there are spotty clusters of drumlins and the intervening landscape is gentle swell-and-swale, like the surface of the ocean. Between mile 387 and Buffalo, the route is underlain by Onondaga limestone, at the base of the Devonian rock section of New York. Large cuts through this unit are present on both sides of the highway at mile 387.5. At Buffalo, near the junction with the Interstate 290 bypass, there is a huge quarry in the Onondaga, where it is used for crushed stone and to make soda ash ($NaCO_3$) from common salt by the Solvay process.

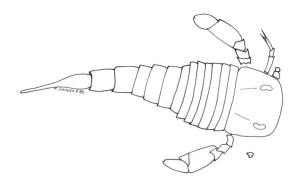

Sketch of fossil eurypterus remipes *from the Silurian Salina group. In 1984, the New York Legislature named this the state fossil. An ancestor of the modern horseshoe crab, this creature sometimes grew to nine feet long!*
Drawing by T. Jancek

Interstate 90:
Buffalo—Pennsylvania State Line
73 mi./118 km.

This stretch of Interstate 90 closely follows the Lake Erie shore. As in the case of the Lake Iroquois predecessor to Lake Ontario, Lake Erie also had its own versions of proglacial lake stages, called Lake Wittlesey and Lake Warren, whose shorelines extended far beyond the present lake. The shorelines are delineated by preserved beach strands which appear as narrow, sandy shelves more or less parallel to the present shore. Most have been dissected by modern erosion, but erosional remnants can be correlated by their similar elevations. Most of the way, the highway is narrowly sandwiched between these ice age strands and the present one.

From the standpoint of physiography and bedrock, the route follows the Portage scarp at the edge of the Erie Plain (see US 20: Buffalo—Canandaigua discussion). The scarp, which is also the northern boundary of the Allegheny Plateau, is on weak shales interlayered with resistant sandstones, rather than a single cliff-forming caprock; therefore it appears as a steep slope rather than a cliff.

Between Exits 50 and 53, the route passes over flat country developed on Onondaga limestone, the lowest formation in the Devonian stratigraphic section of western New York. This unit is extensively quarried in Erie County for crushed stone and used in the Solvay process to make sodium carbonate. Contact with the base of the overlying Hamilton group is passed just south of Exit 53, where bedrock is exposed in the banks of Cayuga Creek. Brownish, rusty

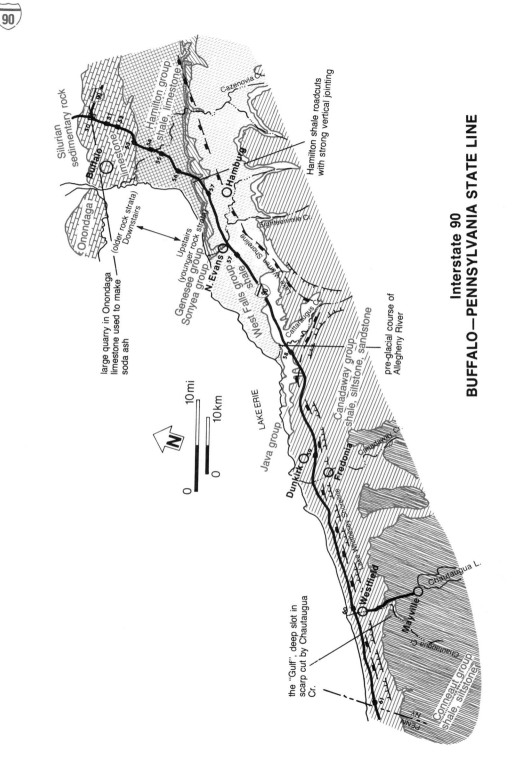

Interstate 90
BUFFALO—PENNSYLVANIA STATE LINE

Silurian sedimentary rock

Cazenovia Cr.

Hamilton group shale, limestone

Hamilton shale roadcuts with strong vertical jointing

Buffalo

limestone?

Onondaga

large quarry in Onondaga limestone used to make soda ash

(older rock strata) Downstairs

Upstairs (younger rock strata)

Hamburg

Eighteenmile Cr.

Genesee group

Sonyea group

West Falls group

N. Evans

shale

Warm Shoreline

Cattaraugus Cr.

Canadaway group shale, siltstone, sandstone

pre-glacial course of Allegheny River

Java group

LAKE ERIE

Dunkirk

Fredonia

Canadaway Cr.

Silver Cr.

Westfield

Mayville

Chautauqua L.

Conneaut group shale, siltstone

Chautauqua Cr.

the "Gulf", deep slot in scarp cut by Chautauqua Cr.

PENN
N.Y.

N

0   10mi

0   10km

shale with strong vertical jointing appears in cuts between Exits 56 and 57. The road continues to climb rapidly up the "staircase" of Devonian strata between Exits 56-58, crossing, in order, the Genesee, Sonyea, West Falls, and Java groups. Good exposures are visible in the banks of several entrenched streams along the way. The streams, including Eighteenmile Creek, and Cattaraugus Creek, have cut down in response to lowering of their base level of erosion from the elevation of Lake Warren to that of Lake Erie.

Cattaraugus Creek lies along the course of the preglacial Allegheny River. As a result of glaciation, the Allegheny is now a completely separate system from Cattaraugus Creek which, near the village of Gowanda, has cut an impressive postglacial gorge called Zoar Valley into the Allegheny Plateau.

Between Exit 58 and the Pennsylvania border, the road follows the outcrop band of the Devonian Canadaway group, which is exposed in a few roadcuts of brownish shales, siltstones, and sandstones. The scarp is nearby all the way, and especially open to view between miles 475 and 476 (large mile signs). Between Westfield (Exit 60) and Mayville, 7 miles to the southeast, the scarp rises 800 feet. Chautauqua Creek, which passes through Westfield to Lake Erie, has cut a deep slot into the edge of the scarp, designated "The Gulf" on the latest 7.5-minute topographic map, but locally called Chautauqua Gorge.

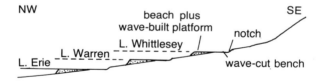

Schematic cross-section through shoreline features of glacial lakes Whittlesey and Warren, and modern Lake Erie.

See map pages 188 and 189.

US 20:
Buffalo—Canandaigua
81 mi./131 km.

THE IROQUOIS TRAIL

Between Buffalo and Depew, actually on NY 130, the route traverses the featureless Erie Plain, on which the small amount of relief mostly results from glacial deposits. The Onondaga scarp, not very impressive in this part of New York, passes through the town of Williamsville about 5 miles to the north. East of Depew, surface relief and elevation increase over middle Devonian shales and limestones, and the gentle rise of the Portage scarp comes into view in the south. The Portage, which marks the northern limit of the Allegheny Plateau, differs from the Onondaga and Niagara scarps in that it is not carved from a single discrete caprock layer. Instead, it is developed on a succession of weak upper Devonian shales with interlayered durable sandstone beds, so that it appears at various stratigraphic levels. Where it is well-developed, however, the scarp attains a height of 900-1000 feet. Because of its structural make-up, it is subject to piecemeal destruction. Its trend on the surface is rather irregular, and it everywhere appears as a steep slope, rather than a cliff, with an incline adjusted to its peculiar makeup of hard and soft layers (see Niagara Frontier section for map).

Between Buffalo and the village of Alden (20 miles), the road climbs 260 feet; it climbs 200 feet higher in the next 6 miles to Darien Center as it mounts the lower slopes of the Portage scarp. Much of the elevation and hummocky relief from Alden eastward, however, results from glacial deposition in the so-called Alden moraine, part of an intermittent moraine belt that extends eastward at least to Syracuse, and skirts the northern ends of the Finger Lakes East. Between Alden and Pavillion Center (NY 19 intersection), the road goes up and down like a rollercoaster over ledges developed on the upper Devonian Genesee group strata, notched by numerous north-flowing small streams. Over most of the way these ledges end in low, but persistent, scarps a mile or two north of the highway. The Portage scarp really begins in the sandier, more resistant, upper Devonian beds a few miles south of US 20, and may be seen to best advantage on US 20A, parallel to this road about 10 miles to the south.

In the 15 miles between Pavillion Center and Avon, the road drops some 400 feet off of the lower slopes of the Portage scarp into middle Devonian shales of relatively low erosional resistance. Again, the road goes up and down like a rollercoaster, but here it is because it crosses part of the great drumlin field of western New York. The drumlins here are not as high as most of those farther north, but they are well-formed and clearly visible under plowed fields and pastureland. The effect is an esthetically pleasing, streamlined, undulating landscape, great for cross-country skiing. The surface flattens rather abruptly 5 miles west of Avon.

Just east of Avon, you cross the Genesee River, the only river to cross New York from one side to the other. Beginning in the hills of Potter County just over the state line in Pennsylvania, the river goes through several moods in its traverse across the land to Lake Ontario. Avon lies in the middle of a 60-mile stretch between the Letchworth and Rochester gorges, where the river assumes its laziest character: a very underfit stream in a broad floodplain that must have been built when the river was much fuller and larger. In its present state, the Genesee wanders freely over this wide plain in great, looping meanders. Flooding is now largely controlled by the Mt. Morris dam at the northern end of the Letchworth Gorge in Letchworth State Park. There are, however, numerous oxbow lakes here in the floodplain, formed where meanders became so convoluted as to nearly loop back on themselves, and then were cut off and abandoned by divide-breaching floodwaters. The broad floodplain and wide and relatively deep bedrock valley in this stretch of river owe their existence to the preglacial Dansville River, which had quite a different course and discharge from the modern river. From Mt. Morris to here, and extending a few miles northward to West Rush, the modern river chose to re-occupy the old Dansville Valley. South of Mt. Morris, however, it did not do so, and instead has carved a new channel into the Letchworth plateau. The old, broad, deep channel, which continues for 20 miles southeastward from Mt. Morris to Dansville, now contains miniscule Canaseraga Creek. This is only part of the story, of course, but generally speaking, glacially-induced drainage changes like this are the rule, rather than the exception, in western New York.

Avon is also at the juncture of Conesus Creek, which drains the westernmost major Finger Lake, Conesus Lake, 5 miles southeast of the village. Between Avon and Lima (8 miles), the topography is once again marked by drumlins, and the road weaves slightly to avoid cutting through the hills. One of these, with a microwave tower on it, is beautifully profiled when viewed from the west, with telltale steep, north slope and long, tapering, southern tail.

Between Lima and West Bloomfield (4 miles), you pass over Honeoye Creek, here entrenched about 100 feet into middle Devonian shales. The creek carries the outflow from Hemlock, Canadice, and Honeoye finger lakes. Beginning less than a mile south of the highway, there is an extensive pitted kame-moraine field replete with mounds of morainal till, kames left by meltwater stream-washed sediments dropped in ponds adjacent to the ice, kettle holes and lakes, and eskers. The latter are long, winding, narrow ridges of sand and gravel that represent sediments deposited along streams which flowed in tunnels under the glacier. Another well-known glacial complex like this is a few miles directly north of here at Mendon Ponds County Park. Still another, called the Pinnacle Hills, lies farther north at the south side of Rochester. Each represents a relatively stationary position of the ice front as it gradually receded northward during the closing stages of Wisconsin glaciation, the last Ice Age. The deposits farther north, of course, are the most recent.

Farther downstream, beyond Honeoye Falls, Honeoye Creek flows westward along the broad, pre-Wisconsin channel of the Dansville River before joining the Genesee near West Rush. The ancestral Dansville River flowed eastward from West Rush for 13 miles to Fisher, then turned northward to Irondequoit Bay. The bay is the drowned mouth of the old river. The modern Genesee flows directly north from West Rush through Rochester to Lake Ontario.

From West Bloomfield to Canandaigua (13 miles), the road continues up and down over beautiful, rolling drumlin country, with occasional panoramic views to the south. Strangely, the drumlins are numerous and well-formed along route 20 and south of it between Canandaigua and Pavillion Center, but are virtually absent in a 5-mile strip north of the highway. The densest and most distinctive drumlin field begins north of this 5-mile strip and extends to Lake Ontario. At least part of the explanation lies in the slope of the land. US 20 is routed along the lower slopes of the Portage scarp; the land to the north is relatively flat, and the slope steepens gradually south-ward. The upslope advance of the glacier must have created just the right conditions of pressure and flow at the base of the ice to mold these peculiar hills. This rise in the landscape to the south is clearly visible from the summits of some drumlins near Canandaigua. Entering Canandaigua from the west, the road drops some 300 feet into the lake trough.

Slump structures in glacial outwash sediments

See map pages 188 and 189.

US 20:
Canandaigua—La Fayette
70 mi./113 km.

This scenic route touches the northern tips of three of the largest Finger Lakes: Canandaigua, Seneca, and Skaneateles, thus affording excellent views southward over the water. Part of it lies within the great drumlin field of western New York with its lovely rolling countryside. The Onondaga scarp is practically non-existent over most of the way, not because of its lesser resistance, but apparently because glacial scour in this region was more intense than to the east or west.

At Canandaigua Village, the road passes close to the north shore of the lake by Kershaw Park, and crosses the lake outlet near the rise up the east side of the trough. All of the Finger Lakes drain northward to Lake Ontario because the Valley Heads moraine dams their southern ends. Canandaigua and all of the Finger Lakes to the east, including Keuka, Seneca, Cayuga, Owasco, Skaneateles, Otisco, Onondaga, and Oneida (the last two not part of the Finger Lakes proper, but finger lakes nevertheless), drain into the Oswego River which, in turn, empties into Lake Ontario at Oswego. In general, this coalescing network drains eastward through a lowland belt developed on weak Silurian Salina group rocks, consisting of such easily-eroded materials as rock salt, gypsum, shale, and dolostone. Keuka Lake differs from the rest in being considerably farther south and draining across the intertrough divide from Penn Yan to the middle of Seneca Lake. The New York Barge Canal follows this natural lowlands water course from Lyons, north of Geneva, to Oneida Lake with four locks and several channel modifications along the way. North of this lowland belt is an east-west drainage divide developed on Lockport dolostone, the same resistant formation as the caprock of Niagara Falls, and all that is left of the Niagara scarp.

For the Finger Lakes West, all lakes west of Seneca, the northward drainage here is a reversal of the preglacial drainage direction. In the Finger Lakes East, the apparent preglacial drainage was northward as it is now, but it has been modified by glaciation.

Between the Canandaigua outlet and Geneva (17 miles), the road climbs about 200 feet over the intertrough divide. What little relief you see along the way is due chiefly to low drumlins, not to bedrock erosion. The intertrough divides along the route are not very high or dramatic west of Skaneateles, because the upland surface gradually descends northward and blends into the lowlands. Troughs deepen and become more fjord-like southward as they slice into the Allegheny Plateau, and the upland surface there locally rises 1500 feet between some troughs.

On the west side of Geneva, the highway descends about 450 feet into the Seneca Lake trough, with open views over the water near the city. The lake surface is 242 feet lower than that of Canandaigua Lake. The northern end of the lake is almost rectangular, apparently because it is dammed by a cross-trending moraine but much lower and less dramatic than the Valley Heads moraine at the south end. This is part of the intermittent east-west moraine belt discussed in the Buffalo-Canandaigua section of US 20, and encountered at Alden east of Buffalo. The tell-tale hummocky, and somewhat pitted, moraine topography occurs on the north side of Geneva.

The route east of Geneva hugs the north shore with wide open views of this largest of the Finger Lakes. With an astounding, maximum depth of 633 feet, it is also the deepest of all, nearly 200 feet deeper than Cayuga, the next deepest. Even these incredible depths don't tell the whole story of the troughs, for the bedrock floors lie an untold distance farther down, buried under great depths of glacial, stream, and lake sediments. The extent of glacial erosion of the bedrock in this region, both in widening and deepening the original stream valleys, has been truly phenomenal. This has not, however, resulted from just one glacial episode, but from at least four widely separated advances, beginning about 2 million years ago. Each succeeding glaciation carried on the process of gouging, scouring, scraping, and smoothing. But it is now widely believed that most of the carving of the troughs was accomplished during the first glaciation when the angular, stream-sculptured landscape offered the greatest resistance to ice movement. By contrast, later ice sheets slid more

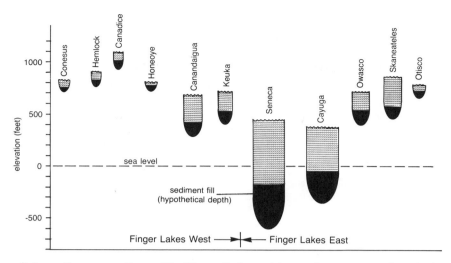

Schematic cross-sections of the Finger Lakes, with greatly exaggerated vertical scale. Each shows actual maximum depth of water, but depth of sediment fill is unknown.

easily over a streamlined surface and expended much less energy in modifying it. The north-south troughs were more deeply eroded than the upland because the ice piled into them and formed thick tongues, but the upland surface has also been reduced an estimated 200 feet more or less.

The road follows the Seneca Lake outlet between the northeast corner of the lake and the Montezuma Marshes at the north end of Cayuga Lake (15 miles). The stream course has been extensively channelized and deepened as part of the Cayuga and Seneca Canal, a feeder to the State Barge Canal system, which it joins near Montezuma. Along the way, the canal passes through Waterloo, Seneca Falls, and the Montezuma Marshes, and descends 71 feet through four locks.

US 20 passes right through a heavily-wooded section of the Montezuma Marshes. This extensive region of bogs and ponds is now under federal protection as the Montezuma National Wildlife Refuge, encompassing ten square miles of habitat. The region is an extension of the bedrock trough of Cayuga Lake, filled with sediments to unknown depths, as under the lake itself. Cayuga Lake is the only one of the Finger Lakes that extends northward as marshland this way; all of the others square off against moraine. Geologists are undecided as to the reason for this anomaly. The Seneca River, now canal, passes right through this swamp, and joins the Clyde River, carrying the Canandaigua outflow. Much of the swampland has been channelized to accommodate barge traffic from Seneca and Cayuga lakes, and on the main Barge Canal itself. It is not hard to imagine what a miserable job it must have been to dig and line the original Erie Canal with stone in the early part of the last century, when most of the work was done by hand. Picture the scene: a labor force of 2000-3000, many of them Irishmen imported for the purpose, often standing waist deep in water and muck, preyed upon by leeches and hordes of mosquitoes. Laborers by the hundreds fell ill with malaria or pneumonia, and many died. You can lay to rest any idea you may have had that these scourges are restricted to tropical regions.

The Erie Canal was a monumental engineering achievement that immeasureably advanced the economic development of New York in the last century. Its main purpose was to facilitate trade with the American west by providing a continuous waterway between the Hudson River at Albany and Lake Erie at Buffalo, an overland distance of more than 364 miles. This was accomplished in the period from 1817-1825.

The routing of the canal shows, perhaps, better than any other example how close the relationship can be between geology and human development. The topograpy and the drainage established on it are largely controlled by the type and "structure," or geometry of deformation, of the underlying bedrock. The Erie Canal utilized, as fully as possible, the existing east-west waterways, such as the Mohawk River, Oneida Lake, Seneca River, Clyde River, and others, and followed their lowland belt to minimize topographic obstacles and cost, while, at the same time, bridging the vertical distance of 545 feet between Lake Erie and the Hudson River at Albany. If a more northerly or southerly route had been chosen, the canal might never

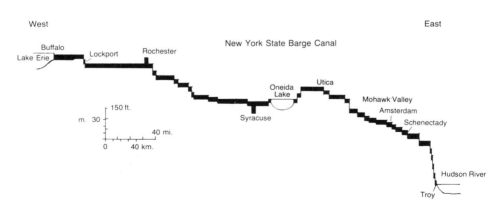

Cross-section through the New York State Barge Canal (Erie Canal), with greatly exaggerated vertical scale. Each step represents a canal lock.

have been finished. After its completion in 1925, and owing to its tremendous commercial success, numerous connecting canal systems were built, including several "lateral" canals that cut across the geological grain. One of the lateral type was the Genesee Valley Canal between Rochester and the Allegheny River south of Olean, which required 112 locks to cover a distance of only 124 miles! The beginning of the end for the canal era came in the 1860s, when railroads were able to provide cheaper and faster transportation.

East of the Cayuga and Seneca Canal on the eastern side of the Montezuma Marshes, US 20 cuts through the southern edge of the drumlin field with exceedingly fine, slender drumlins that display a length:width ratio of up to 6:1, pronounced, longitudinal asymmetry with steep north-northwest-pointing noses and gentle south-southeast-trailing slopes, and heights of 100 feet or more. Their abundance and strikingly linear form demonstrates, once again, the extreme sensitivity of these unique features to slight changes in slope under the ice sheet. The glacier here climbed diagonally out of the marshes part of the Cayuga trough onto the Cayuga-Owasco divide, which now stands only a little over 100 feet above the marshes level. Emphasize "now," for then it emerged from the bedrock trough. Drumlins immediately north and east of this upslope region are much more equidimensional in outline, more symmetrical in long profile, and higher, possibly because the ice exerted less downward pressure where it moved against level or downsloping land.

Several roadcuts between the Marshes and Auburn (9 miles), expose the stony till of the drumlins. Near Auburn, the cuts expose the lower Devonian Helderberg group limestones and middle Devonian Onondaga limestone, which become increasingly prominent ridge-formers farther east.

Auburn straddles the Owasco Lake outlet just north of the lake. Morainal deposition in the northern trough extension is apparent, as in the case of Canandaigua and Seneca lakes, in the hummocky topography of the village and environs.

Seven miles east of Auburn, the road passes through Skaneateles along the north shore of Skaneateles Lake, the largest of the high level lakes. Its water surface, at 863 feet above sea level, is third highest of the Finger Lakes, after Canadice (1076 feet) and Hemlock (905 feet), and nearly 480 feet higher than Cayuga Lake.

Skaneateles is certainly one of the gems of the Finger Lakes, especially when viewed from points near its southern end. Here, the closeness and steepness of the trough walls create the impression of a glacial fjord, and the unique, greenish tint of its clear water further enhances the image because it so resembles ocean water in appearance.

The route between Skaneateles and La Fayette (18 miles), called the Cherry Valley Turnpike, offers scenery markedly different and more spectacular than anything farther west on US 20. Here, the road rides high on the dissected plateau well south of the now prominent Onondaga scarp, and it plunges right into the Otisco Lake and Tully Valley troughs.

This is a region of ice-marginal spillover channels, one of which, called Pumpkin Hollow, lies 2-3 miles north and roughly parallel to the highway. Such channels are carved where meltwater in a high proglacial lake drained across the intertrough divide along a relatively stable ice margin into an adjacent trough. Pumpkin Hollow is a particularly interesting one because of its so-called Cedarvale "hanging deltas." Deltas were formed in many places where sidestreams emptied into the proglacial lakes, with their top surfaces corresponding to the lake levels at the time of their sedimentation. But the lakes tended to have several "stands" at successively lower altitudes as the ice receded slowly northward, and new, lower outlets were opened. The old deltas were left "hanging" above the water and, as delta-formation continued at the new lake level, they were trenched by the same streams that built them.

Pumpkin Hollow, which spans the divide between the Otisco and West Onondaga troughs, was carved contemporaneously with the Smoky Hollow spillover channel between East Onondaga and Butternut troughs. Later ice stands resulted in three other, progressively lower, spillover channels north of Smoky Hollow, called Clark Reservation, Rock Cut, and Erie Canal. Clark Reservation, near Jamesville, which is now set aside as a public park, contains some of the most impressive meltwater erosional features in the state. The highest bedrock threshold of this spillway is about 760 feet above sea level, so this channel must have been abandoned when the Onondaga trough lake, called Lake Cardiff, fell below that level. The park contains a large, natural amphitheater where the stream once poured over a hundred-foot-high precipice, and Green Lake now rests in the plunge basin of those former falls.

The Tully trough here is over 1000 feet deep and the panorama from the valley sides over the broad Y-junction with the east and west Onondaga valleys is awesome. The valley floor is about 1.5 miles wide.

See map pages 180 and 181.

US 20:
La Fayette—Guilderland
119 mi./193 km.

THE CHERRY VALLEY TURNPIKE

This section of US 20 is one of New York's most spectacular highways. The route follows the northern edge of the Allegheny Plateau and crosses numerous glacial troughs. There are many open views north and south into them, and unrestricted panoramas from the divides. The troughs are all part of a branching system that coalesces southward to the Susquehanna River. The Valley Heads moraine forms through-valley drainage divides, separating north-flowing streams from south-flowing streams. The moraine barrier dies out quickly east of La Fayette, as it also shifts northward.

The Butternut trough immediately east of La Fayette is certainly the most impressive of all the crossings. The road makes a big, southward loop into the valley, with clear views south to the Valley Heads moraine at Clark Hollow that begins only two miles away. The steep-sided Jones Hill profiled beyond the divide, above Labrador Pond, is a popular take-off place for hang-gliders in summertime. From the top to the campground where they land at the north end of the pond, the drop is only 700 feet; but the gliders often ride the updrafts for an hour or more. Labrador Pond lies at the head of the east branch Tioughnioga River, part of the Susquehanna drainage network.

From near the crest of the hill east of the trough, it is possible to look north over the Jamesville Reservoir to Syracuse and the Ontario Lowlands, visibility permitting. This view expands to a 360-degree panorama at the Village of Pompey, 6 miles east of La Fayette. Note the general accordance, or alignment of the summits in nearly all directions, a feature that is almost certainly inherited from an ancient, nearly level, erosional surface that is now deeply furrowed with stream and glacial passageways.

East of Pompey the road passes over rolling upland country and dives into Pompey Hollow, or Limestone Creek trough, past big roadcuts in dark gray Hamilton shales. Cazenovia Lake, 10 miles east, lies in a depression about 400 feet higher than Pompey Hollow, with the picturesque small town of Cazenovia at its southern tip. The lake drains by the village into Chittenango Creek, which turns immediately northward and flows through a narrow gorge. Beginning about 4 miles downstream, it passes over tiered falls and rapids as it descends the scarp-forming Onondaga and Helderberg limestones in Chittenango Falls State Park, dropping about 300 feet in a little over a mile. In reality, this is a part of the main scarp that has been cut back by the stream, so that the limestone band, in plan view, now makes a long V that points southward.

Between Cazenovia and Morrisville (11 miles), the road traverses rolling upland topography and crosses some shallow, marshy valleys that don't quite qualify as troughs, but which, nevertheless, have been glacially deepened, widened, and sedimented. Much of the relief is step-like, developed on resistant sandstone interlayers in Hamilton shales. Several roadcuts along the way expose dark grayish or

Chittenango Falls north of Cazenovia, cascading over Onondaga limestone

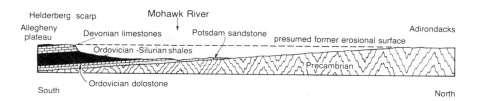

North-south cross-section from the Helderberg scarp to the edge of the Adirondack Mountains, showing a hypothetical former erosional surface now deeply dissected   Courtesy Kendall/ Hunt Publishing Co.

brownish rocks of this group, in which the thicker, sandier interbeds weather in relief. The Morrisville Swamp, north of Morrisville, is the inauspicious spawning ground for the Chenango River, which joins the Susquehanna at Binghamton. The valley becomes much more trough-like south of Morrisville.

Sangerfield, 14 miles from Morrisville, also lies in a swampy, shallow valley, as does Bridgewater, 8 miles farther east. At the time of this writing, there is an excavation in the Bridgewater Flats, north of the village, that exposes glacial varves, those distinctively laminated clay deposits formed by slow sedimentation in glacial meltwater lakes.

Richfield Springs, 13 miles east of Bridgewater, lies at the northern end of Canadarago Lake, one of two small outlying finger lakes in this region. The other is Otsego Lake, a few miles farther east. Unlike the major Finger Lakes to the west, each drains southward into the Susquehanna River. The historic village of Cooperstown, the "Village of Museums," with its National Baseball Hall of Fame, lies at the southern tip of Otsego Lake.

The 34 miles between Richfield Springs and Sloansville offers beautiful open scenery, where the road skirts the Onondaga-Helderberg scarp. Near Sloansville, it descends to the Ordovician lowlands of the Mohawk Valley. Over much of the way, the highway dips and climbs in crossing several troughs that crease the northern edge of the Allegheny Plateau. Panoramic views north over the Mohawk Lowlands begin just east of Springfield (6 miles), located at the intersection with New York 80, and get progressively better. Perhaps the best of all is at the rest stop located 8.5 miles east of Springfield. Here, the outcrop bands of resistant Onondaga and Helderberg limestones narrow down and form a particularly high, steep slope. The overlook, perched near the top, is like a grandstand seat on the 50-yard line. The escarpment is profiled to the east and west. The Mohawk River is only 9 miles distant. Toward the northeast it has cut deeply into the upthrown block of the Noses fault, the same one that forms the northwest side of the Sacandaga Lake depression farther north. Across the valley, the Adirondacks and Tug Hill appear only as broad, gentle swells on the horizon, and are not very impressive; but the immensity of the scene is itself a source of wonder.

The large new roadcuts on the hillslope just west of the overlook are in Onondaga limestone, with some shale interbeds. The broad erosional steps, so conspicuous on the slopes immediately north of the highway, are developed on the more resistant layers in this unit and on top of the underlying Helderberg limestones. Eastward of the overlook there are several roadcuts in Helderberg limestones, where the road descends the stepped slope through progressively lower stratigraphic horizons. The road crosses the base of the Helderberg limestones 14 miles east of the overlook; the underlying step is on the Silurian Cobleskill limestone, beyond which the road drops to the Ordovician shales of the lowlands. There are numerous drumlins north of Carlisle.

Between Sloansville and Esperance (3.7 miles), the road dips into the Schoharie Valley. With its headwaters in the Catskills, Schoharie Creek is the only major river in the eastern part of the Allegheny Plateau that flows northward across the Helderberg scarp, joining the Mohawk River at Fort Hunter. The valley was extensively scoured during the Ice Age and filled with sediments during glacial recession. At one stage, a long slender lake, called Lake Schoharie, occupied the section of valley for several miles southward from a moraine at Esperance, and that section of valley now is especially broad and flat-floored as a result of lake sedimentation.

Between Esperance and Gifford (11 miles), the highway goes up and down over a landscape deeply grooved and topped with crowded low drumlins left by ice that pushed westward up the Mohawk Valley from the Albany region.

On the next 8 miles between Gifford and Guilderland, there are several excellent viewpoints. To the south, the Helderberg scarp has progressed from a stepped slope to a single, discrete barrier over 120 feet high, as it also curves around to the south to form the eastern boundary of the Allegheny Plateau next to the Hudson Lowlands. This high rampart is referred to as the Helderberg Mountains. One of New York's premier parks, John Boyd Thacher State Park, is perched atop the cliff. From lookouts at the scarp rim, it is possible, in one sweeping view, to see the Adirondacks, the Green Mountains of Vermont, and the Taconics of Massachusetts. The broad expanse of flats around Albany, 13 miles to the east, is the sand plain of Lake Albany, a large meltwater lake that occupied the valley during recession of the Wisconsin ice sheet. The sand, which is generally less than 50 feet thick, blankets an erosional surface developed on bedrock

during the Tertiary, or pre-Ice Age, period. As in other large meltwater lakes, like Lake Iroquois or Lake Vermont, there are many preserved shoreline features that were formed around Lake Albany, such as beach strands and deltas. The Schenectady Delta, for example, is an immense delta formed during the early lake stages when the ice margin was not far north. The Mohawk then carried all of the drainage from Lake Iroquois and, in fact, nearly all of that of the proglacial Great Lakes, and thus carried copious quantities of sediment into Lake Albany.

John Boyd Thacher State Park and the Helderberg scarp; weak Snake Hill shale (Ordovician) in foreground   Courtesy Kendall / Hunt Publishing Co.

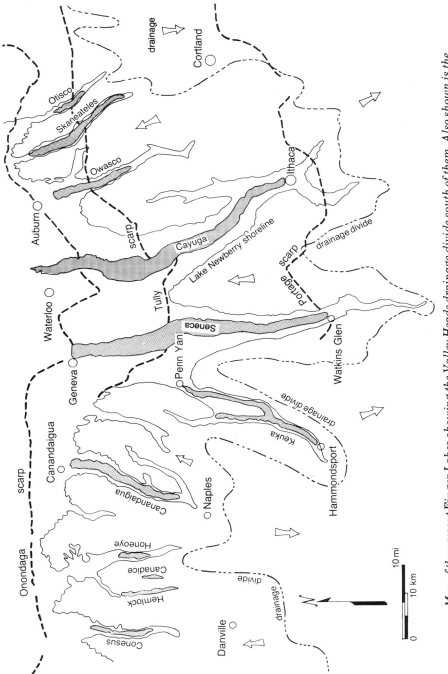

Map of the present *Finger Lakes* showing the *Valley Heads* drainage divide south of them. Also shown is the shoreline of the earlier interconnected glacial meltwater *Lake Newberry* and the lines of three principal scarps that cross the region from east to west.   Courtesy Kendall / Hunt Publishing Co.

IX

THE FINGER LAKES REGION

The Finger Lakes comprise one of New York's most distinctive scenic areas. The lakes occupy generally north-south-trending, preglacial, stream valleys that were deepened and widened by glacial gouging during the Ice Age, and then dammed up by the so-called Valley Heads moraine. The formation of the lakes is exemplified in the following blow-by-blow account of Cayuga Lake.

Imagine the scene around Ithaca at Valley Heads time during Wisconsin deglaciation. The ice front began slowing its northward retreat when it reached Spencer, perhaps in response to a temporary climatic cooling, and continued at a very slow rate until it got to Ithaca. There was enough time for much till to melt out of its ice-locked prison and pile up on the valley floor. Since it was unevenly distributed in the ice, it formed uneven piles.

As is usual in wasting glaciers, many blocks of stagnant ice broke off and some were buried in the accumulating sediment. Much later, when these melted, they left pits, called kettles; today some pits remain filled with water, known as kettle lakes.

Meltwater ponds and lakes filled hollows in the ice and alongside the ice against the valley sides. In some, gravelly kame deltas were deposited by inflowing streams. Deltas on the ice were later dropped on the moraine when the ice melted, forming more mounds and hummocks. Meltwater washing over these deposits redistributed it as an outwash apron south of the moraine.

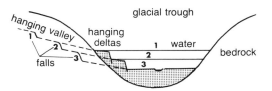

Deltas formed during proglacial lake stages at the mouths of hanging valleys are left hanging themselves when the lake level is lowered, as shown in steps 1-3. In steps 2 and 3, note that the side stream trenches the earlier delta, as waterfalls form and recede upstream.

By the end of Valley Heads time, the ice front stood near Ithaca, and a large lake had begun to form south of it to the height of the moraine. Then, perhaps because the climate began to warm more rapidly, the recession accelerated, and the lake lengthened northward. Since it was confined to the Cayuga trough and banked against the valley-blocking ice on the north, the copious meltwater filling the basin had only one escape—to the south over the moraine. It continued to pour southward through a series of sluiceways to the Susquehanna River which, at this stage of the retreat, was the main outlet for all of the forming Finger Lakes.

Later, when the ice front had receded farther north, lower northern outlets opened up, and the lake level dropped below the moraine crest. A finger lake was made! Other outlets at even lower levels formed later, and the level dropped again, and again. The modern lake, therefore, can be considered just the lowest phase in a succession of lakes that generally got smaller and lower with time. Among other more interesting byproducts are the hanging deltas we find today on the sides of the trough. They are just ordinary deltas like those building into the lake today at the mouths of lateral tributary streams, but they formed earlier at the higher levels of the proglacial lakes. When the level dropped, they were left high and dry. The streams that formed them, however, continued to flow, forming new deltas at the new lake level and cutting into the older, hanging deltas.

These same tributary streams are also the sites of the many glens, perhaps the most famous of the Finger Lakes features. Most, but not all, of the glens are notched hanging valleys, valleys of small, preglacial tributaries to the original Cayuga River that were glacially gouged, but not nearly as much as the main trough. The larger size and more favorable orientation of the main valleys like the Cayuga trough led to their deeper gouging by thick tongues of the advancing ice. When the ice wasted, the side valleys were left hanging high above the bedrock floor of the trough, and any which lay athwart the ice flow were filled with drift. Newly established upland tributaries spilled over the edge, and notching began as the falls wore away bedrock and glacial fill and slowly retreated upstream. Many seem to have new courses that deviate from the pre-Wisconsin ones. Enfield Glen, for example, cuts through bedrock over most of its length; in places, however, the old, drift-filled gorge is exposed in the walls of the new gorge.

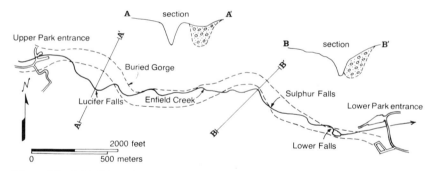

Map of Enfield Glen (Robert H. Treman State Park) showing the outline of the buried pre-Wisconsin gorge in relation to the present one marked by Enfield Creek Adapted from NYSGA Guidebook, 31st annual meeting, Cornell University, 1959. Courtesy Kendall / Hunt Publishing Co.

All of the Finger Lakes now drain northward to Lake Ontario; for most, this represents a reversal of the preglacial flow direction. The Genesee River drainage was also profoundly modified by Wisconsin glaciation. In the preglacial version, the trunk river, called the Dansville River, had its main headwaters in the Canandaigua Lake valley, and at its lower end it passed through Irondequoit Bay. The modern version, instead, has excavated the gorges at Letchworth and Rochester. The whole preglacial system was marked by barbed junctions indicative of stream piracy. In re-establishing itself after the Wisconsin ice sheet withdrew, the modern Genesee had no difficulty in finding its old route from Pennsylvania to Portageville. From here, however, the old, northeast channel to Sonyea was blocked by the Portageville moraine, a segment of the Valley Heads moraine, and the river was forced to go north instead over the Letchworth plateau.

Meanwhile, the old Dansville River route was even more plugged up by Valley Heads moraine at the south end of Canandaigua Lake, and between Dansville and North Cohocton. These blockages prevented the entire western branch from being re-established as part of the new system. The Dansville-Mt. Morris trough and the Nunda-Sonyea Valley now contain tiny creeks much too small for their valleys in place of the former large river branches. Downstream, the new river was prevented by other drift blockages near Avon from reaching its channel to the east and instead continued north across the Rochester plain.

This region includes a 20-mile wide span of the great drumlin field of western New York. It seems possible that the deep ice scour here actually created the conditions favorable for the later molding of drumlins by providing just the right angle of slope opposed to ice flow. Drumlins form by remolding of glacial drift under the moving ice where pressure conditions are just right, meaning that the thickness of the ice is just right, and where there is sufficient meltwater at the base of the ice. A glacier normally gets thinner and moves more slowly toward its terminus as the rate of melting increases. At a certain distance back, therefore, the weight of the ice, rate of flow, and amount of water at the base must be optimum for drumlins to form. The slope angle over which the ice moves can serve as a pressure modifier, increasing pressure where the ice moves uphill and decreasing the pressure where it moves downhill. It's interesting to note that the southern margin of the drumlin field approximatey parallels the Valley Heads moraine and lies about 30 miles north of it. The well-defined northern margin lies some 30 miles yet farther north. One may assume that, beyond the northern limit, the weight of the ice was too great for drumlins to form, and the existing drift was simply smeared or bulldozed southward.

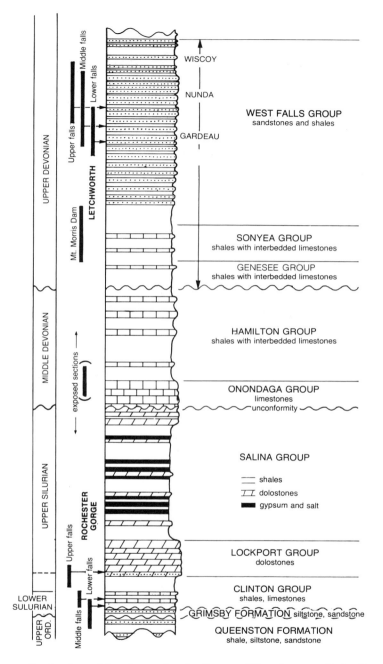

Composite stratigraphic column for the Genesee River between Lake Ontario (base of column) and Portageville at the upper end of Letchworth Gorge, with the falls sections indicated by black bars on the left side.
Courtesy Kendall / Hunt Publishing Co.

215

The accompanying, composite stratigraphic column shows the rock record of the Genesee River country from Rochester to Portageville. As the name implies, the column is a composite of several exposed sections from different parts of the region and, of course, no single exposure shows the complete section. The units are stacked in their proper chronologic order with the oldest at the bottom. Note that the rock section of the Rochester Gorge is similar to that of the Niagara Gorge and comprises the oldest rocks in the region. Note also that the rocks get younger toward the south to the Letchworth Gorge, a product of the southward dip of the beds and rise in elevation shown in the block diagram.

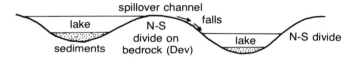

Spillover channels are cut across divides when a high proglacial lake in one trough overtops a divide and water drains across to a lower lake.

NY 89:
US 20—Ithaca
41 mi./66 km.

THE CAYUGA WINE TRAIL

This scenic highway follows the west side of Cayuga Lake from its junction with US 20 by the Montezuma Marshes at the northern end to Ithaca at the southern end. The marshes, which are now under federal protection as the Montezuma National Wildlife Refuge, lie in the sediment-filled northern extension of the Cayuga trough. This is the only one of the Finger Lakes troughs that extends northward beyond the Allegheny Plateau, and geologists are still uncertain as to why it does so (see discussion in US 20: Canandaigua—La Fayette). The Erie Canal and the Cayuga and Seneca Canal pass through this swamp, and the New York Thruway and US 20 cross over it.

The highway crosses the Cayuga and Seneca Canal just south of the US 20 junction, and right about at the southern margin of the drumlin field. The canal is the much-channelized and deepened Seneca Lake outlet that permits barge and pleasure boat traffic between the two lakes, and connects with the Barge Canal at the north end of the marshes. The drumlin field extends a little farther south on the east side of the lake, where the drumlins are strikingly linear and unusually well-formed.

The slopes descending to the lake throughout are dotted with vineyards where some of the East's finest wines are vinted. A key factor that makes this region so well suited for growing grapes is the

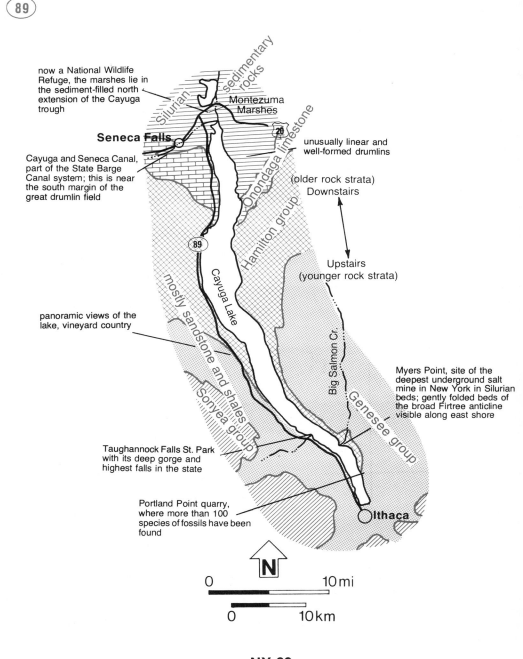

now a National Wildlife Refuge, the marshes lie in the sediment-filled north extension of the Cayuga trough

sedimentary rocks

Silurian

Montezuma Marshes

Seneca Falls

20

Cayuga and Seneca Canal, part of the State Barge Canal system; this is near the south margin of the great drumlin field

unusually linear and well-formed drumlins

Onondaga limestone

(older rock strata) Downstairs

Hamilton group

89

Upstairs (younger rock strata)

Cayuga Lake

panoramic views of the lake, vineyard country

mostly sandstone and shales

Big Salmon Cr.

Myers Point, site of the deepest underground salt mine in New York in Silurian beds; gently folded beds of the broad Firtree anticline visible along east shore

Sonyea group

Genesee group

Taughannock Falls St. Park with its deep gorge and highest falls in the state

Portland Point quarry, where more than 100 species of fossils have been found

Ithaca

N

0 _____ 10 mi

0 _____ 10 km

**NY 89
US 20—ITHACA**

Taughannock Falls, highest in the state; key formations labeled, all belong in the Genesee group.

shaley bedrock, which breaks up into small chips that form a boxwork in the soil. This provides a lot of passageways so the soil is well-drained. The slope angle is also important in draining the soil. Very few of the vineyards are perched on the flat crest of the intertrough divide; most are nearer the lake where the incline is just right. There appears to be none on the steepest slopes at lakeside.

Between the Montezuma Marshes and Taughannock Falls State Park (32 miles), the road lies partly along the shore and partly high upon the slope where you have spectacular, panoramic views over the lake. En route, you cross numerous gullies cut in the trough wall by small, postglacial creeks. Taughannock Creek is a much larger stream that has sliced a lovely canyon with Taughannock Falls at its head. Taughannock Falls are 215 feet high, the highest in the state. Taughannock Point is a modern, fan-shaped delta in the lake built of sediments derived from the cutting of the gorge. The strata exposed in the walls of the gorge are slightly older than those of Buttermilk Glen or Enfield Glen south of Ithaca. This is because the regional dip, or slope, of the beds is slightly southward; along a certain horizon such as the lake level, the beds get older toward the north as you move "down" in the stratigraphic section.

At the entrance to the gorge, there is a small falls sustained by Tully limestone, which is sandwiched between the weaker shales of the Genesee group above and the Hamilton group below. The main falls are about one mile into the gorge and may be reached by an easy trail. Over most of the way the streambed is the broad, flat surface of the Tully, swept clean of the overlying shales by floodwaters. The surface is criss-crossed with joints and pocked with small solution pits. The walls heighten progressively into the gorge, reaching a maximum of 400 feet in the amphitheater of the main falls. The falls are held up by the moderately resistant Sherburne siltstone of the Genesee group, overlying the upper 90 feet of black Geneseo shales. One hundred fifty feet of block jointed Sherburne is capped by 50 feet of Ithaca shales beginning about 25 feet above the falls. The carving of the great amphitheater is mostly due to the action of spray in accelerating rock weathering and erosion. Also the gorge is probably lengthening more slowly since the harder Sherburne beds were intercepted, giving more time for hollowing of the amphitheater. In the early stages, the top of the gorge was below the base of the Sherburne, and there were probably steep rapids in place of falls. It was only recently, geologically speaking, that the Sherburne was intersected and the cataract initiated. The falls recession is slow by human standards, but geologically very rapid. The last major recorded rockfall from the lip of the falls occurred between 1888 and 1892!

On the southern part of the route between Taughannock State

Park and Ithaca (9 miles), you have good views of Myers Point on the opposite shore, the modern lake delta of Salmon Creek, which forms one of the best-known barbed junctions in the eastern Finger Lakes (see discussion of barbed junctions and stream piracy in NY 17: Harriman—Binghamton). The preglacial Salmon Creek drained southward over the plateau; but it was eventually captured by the dominant north-flowing Cayuga River, whose widened and deepened valley now contains the lake. The present Salmon Creek course approximates the preglacial one, preserving the fossil barbed junction.

Myers Point is presently the site of the largest underground salt mine in the Finger Lakes region. Salt is closely assciated with gypsum beds in the late Silurian Salina group. Over 10,000 square miles of western and central New York are underlain by Silurian rock salt; and its extraction has been an important part of New York's economy for a long time. Salt springs near Onondaga Lake were known to the Indians; and trading for salt, evaporated from these springs, is largely responsible for the location of the city of Syracuse.

The salt and gypsum are "evaporite" deposits formed in a Dead Sea or Great Salt Lake type of environment which typified much of the late Silurian period in New York and Michigan. Under these conditions, seawater, or even so-called "fresh" water, is evaporated from shallow basins until their dissolved salts become so concentrated that they precipitate. With just the right balance between evaporation rate and rate of influx of new water, the critical concentration can be maintained for long periods, and thick deposits of only one kind of salt accumulate. At other levels of concentration other salts like gypsum would precipitate. One of the most awesome occupants of these very salty Silurian seas were the eurypterids, otherwise called sea scorpions, an ancestor of the modern king crab that sometimes grew to 9 feet long! The Bertie formation limestones of New York are well known for their abundant, well-preserved fossils of these bizarre arthropods.

The salt beds under Myers Point, along with the overlying and underlying strata, have been upfolded into the broad, gentle Firtree Point anticline, which can be seen along the shore south of the point. The anticline has been drilled to the early Devonian Oriskany sandstone in search of gas, but the amount found was too small to be marketable. The Oriskany is a major producer in other parts of the state.

The Portland Point quarry, a mile southeast of Myers Point, is a famous fossil locality, where over 100 species of corals, bryozoa, crinoids, brachiopods, pelecypods, gastropods, and cephalopods have been found. The overlying Tully limestone has been taken for cement manufacture, riprap, and road stone.

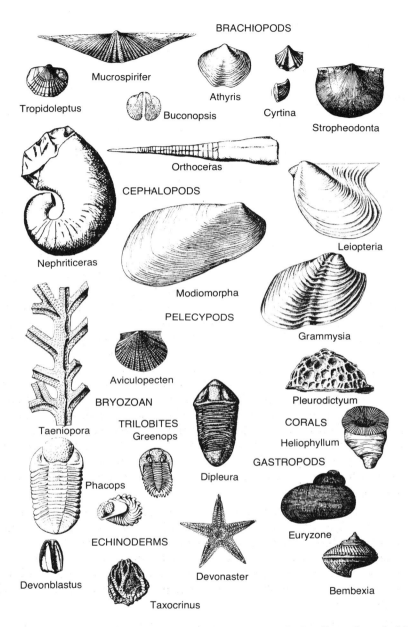

BRACHIOPODS

Mucrospirifer

Tropidoleptus

Athyris

Buconopsis

Cyrtina

Stropheodonta

Orthoceras

CEPHALOPODS

Leiopteria

Nephriticeras

Modiomorpha

PELECYPODS

Grammysia

Aviculopecten

BRYOZOAN

Pleurodictyum

Taeniopora

TRILOBITES
Greenops

CORALS

Heliophyllum

Phacops

Dipleura

GASTROPODS

Devonblastus

ECHINODERMS

Euryzone

Devonaster

Taxocrinus

Bembexia

Representative fossils of the Hamilton group of New York, all at about half scale. Courtesy Kendall / Hunt Publishing Co.

Near Ithaca, the highway descends to the flat delta plain of the south end of the lake, on which much of the city is built (see NY 13: Cortland—Horseheads discussion).

Septarian concretion in Geneseo (Genesee group) black shales of the Hubbard quarry, 9 miles north of Toughannock Point on NY 89.
Courtesy Kendall / Hunt Publishing Co.

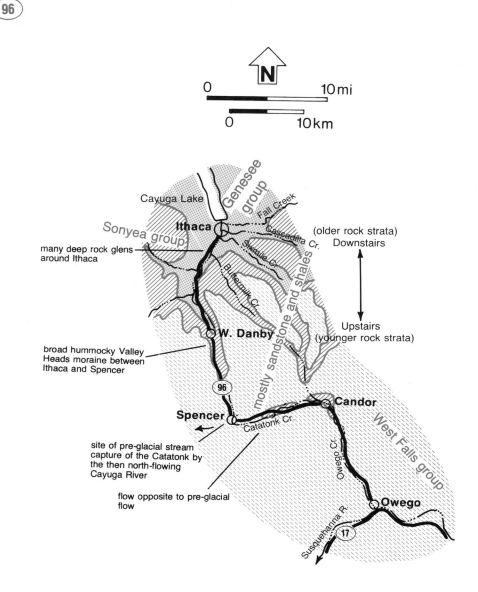

N

0   10 mi

0   10 km

Cayuga Lake

Genesee group

Fall Creek

Sonyea group

Ithaca

Cascadilla Cr.

(older rock strata)
Downstairs

many deep rock glens around Ithaca

Sixmile Cr.

Buttermilk Cr.

mostly sandstone and shales

Upstairs
(younger rock strata)

W. Danby

broad hummocky Valley Heads moraine between Ithaca and Spencer

96

Candor

West Falls group

Spencer

Catatonk Cr.

Owego Cr.

site of pre-glacial stream capture of the Catatonk by the then north-flowing Cayuga River

flow opposite to pre-glacial flow

Owego

Susquehanna R.

17

NY 96
ITHACA—OWEGO

Sketch of an aerial view looking south over the hummocky topography of Valley Heads moraine between Ithaca and Spencer. NY 96B is the dark line on the right.
Drawing by T. Jancek

NY 96:
Ithaca—Owego
38 mi./72 km.

Between Ithaca and Spencer (19 miles), this route crosses over the most remarkable expanse of glacial moraine in New York, the part of the Valley Heads moraine that blocks the southern reach of the Cayuga Lake trough. No other segment of the Valley Heads is as extensive as this one, and none displays the characteristic boulder-strewn, mounded and pitted, kettle-pocked topography as well.

At its highest point a few miles north of Spencer, the moraine is over 1000 feet above sea level, over 600 feet higher than the 389 foot level of the present lake. You can see that this is a pretty effective dam.

The moraine grades to pitted outwash plain between West Danby and Spencer (8 miles), with a lot of marshy hollows, small kettle lakes, and one large one. Kame terraces are also well developed on the sides of the trough in this stretch.

The north-south Cayuga trough meets with the east-west Candor-Van Etten Valley at Spencer in a peculiar T-junction that requires some explanation. Apparently, the preglacial east-west valley was carved along the scarp of the Gardeau formation of the late Devonian West Falls group that now crops out on the hilltops to the south. There are many lesser scarps like this in the Allegheny Plateau region where more resistant beds form caprocks and most go east-west because the beds dip southward. Fall Creek valley near

225

Ithaca, for example, lies along the base of the Portage scarp. The T-junction at Spencer resulted when the headwaters of the preglacial, north-flowing Cayuga River extended to here and captured the Catatonk. The Cayuga was a noteworthy pirate that lengthened itself more rapidly than most other streams in the region. Thus it was able to intercept many others and steal some of their discharge. The best-known case is that of Salmon Creek north of Ithaca which forms a barbed junction with the lakeshore (see NY 89: US 20—Ithaca discussion).

Between Spencer and Candor (8 miles), you follow Catatonk Creek downstream to the east. The present flow is opposite to the preglacial flow to the Cayuga River. The valley is broad and flat-floored, with steep bedrock walls, all features common in valleys that were well established in preglacial time.

Between Candor and Owego (11 miles), a much more constricted north-south valley takes the Catatonk to its junction with Owego Creek. The narrowest part is the site of a preglacial divide that was gouged by ice and later widened by torrential meltwaters. Naturally, the preglacial flow north of the divide was opposite to the present flow.

Owego Creek joins the Susquehanna at Owego. It has all the same physical characteristics as the Candor-Van Etten Valley and obviously was a well established preglacial drainage route. Here, however, the flow direction has not been reversed as a result of glaciation.

NY 14:
Geneva—Horseheads
54 mi./87 km.

NY 14 follows the Seneca Lake trough from one end to the other. The lake occupies only the northern 35 miles of the trough, while the southern part, like all the rest of the Finger Lakes, is filled with Valley Heads moraine and other glacial deposits. The route is an excellent one on which to sense the true spectacle, colossal scale, subtle beauty, and geologic meaning of the Finger Lakes (see also NY 21 and 371: Canandaigua—Cohocton discussion).

Between Geneva and Watkins Glen (38 miles) the road goes along the west side of the lake with breathtaking panoramas from the high points. Perhaps better than any of the other lakes, this one displays the distinctive convex, or "rolled," valley sides that steepen toward the lake. Because of this, the near side appears to drop away below you, and you look down on the far side as if from an airplane. The streamlined curves are a product of glacial erosion. Had the glacier been totally confined to the valley, it would have imparted to it a U-shaped profile like those of alpine glaciers; but this wasn't the case. Instead, the erosion was accomplished by overriding ice sheets with much thicker tongues in the troughs; and the banking of creeping ice along the sides and over the tops of the intertrough divides rounded everything off. The results are convex upper slopes—the parts we see above the lake levels and a concave bedrock floor under the lakes, or under sediment-fill elsewhere—the part we don't see. The depth of the Seneca Lake trough is astounding. At 633 feet maximum, the

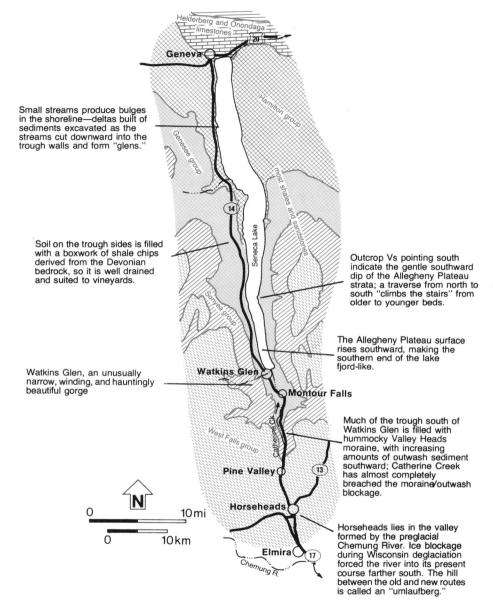

Small streams produce bulges in the shoreline—deltas built of sediments excavated as the streams cut downward into the trough walls and form "glens."

Soil on the trough sides is filled with a boxwork of shale chips derived from the Devonian bedrock, so it is well drained and suited to vineyards.

Watkins Glen, an unusually narrow, winding, and hauntingly beautiful gorge

Outcrop Vs pointing south indicate the gentle southward dip of the Allegheny Plateau strata; a traverse from north to south "climbs the stairs" from older to younger beds.

The Allegheny Plateau surface rises southward, making the southern end of the lake fjord-like.

Much of the trough south of Watkins Glen is filled with hummocky Valley Heads moraine, with increasing amounts of outwash sediment southward; Catherine Creek has almost completely breached the moraine/outwash blockage.

Horseheads lies in the valley formed by the preglacial Chemung River. Ice blockage during Wisconsin deglaciation forced the river into its present course farther south. The hill between the old and new routes is called an "umlaufberg."

NY 14
GENEVA—HORSEHEADS

lake is the deepest of all the Finger Lakes, but this is only part of the story. The bedrock floor is filled to unknown depths with glacial drift and modern sediments.

Along the west side of the lake, you cross several small streams that have carved youthful glens and redistributed the materials as fan-shaped deltas in the lake. Some of the glens, such as Kasong Creek, Big Stream, Rock Stream, and Big Hollow Run, are, in fact, quite large. The fans are best developed in the west side, but some small ones may also be seen on the opposite shore. At Dresden, you also pass the outlet from Keuka Lake, the only Finger Lake that drains directly into another one.

From the high places on the road, note how the upland surface rises to the south. The Seneca-Cayuga divide at the north end is barely 100 feet high, while at the south end it is over 1000 feet high. All of the troughs deepen this way toward the southern ends of their lakes, making those sections look like fjords. In fact, they're formed in much the same way as fjords, but in the final stage are submerged in fresh water instead of seawater.

This is, of course, wine country. The soil that develops on the shaley bedrock generally contains many shale chips that form a permeable boxwork. This feature combined with the slope angles on the trough sides and the relatively dry climate are ideal for vineyards.

Excellent exposures of middle Devonian Hamilton group shales are in the small glens you cross along the northern half of the lake, while the stratigraphically higher Genesee group shales and siltstones are exposed along the southern half, both in the glens and in roadcuts.

Watkins Glen State Park is on the southwest side of the village of Watkins Glen. This is the best known of all the Finger Lakes glens and certainly one of the most beautiful places in eastern North America. Here, Glen Creek has cut a deep and unusually narrow gorge into the thin-bedded Enfield siltstones and shales belonging to the late Devonian Sonyea group that overlies the Genesee group. Because of the relatively uniform resistance of the rocks to erosion, the gorge does not have large falls. Unlike the Enfield and Taughannock gorges, there appears to be very little joint control of the stream course. As a result, the stream twists and turns in a tortuous manner, leaving overhanging walls where meanders have drifted sideways as they cut rapidly downward. This is a showpiece of potholes where, in time of flood, numerous swirling eddies drive sand and gravel that grind hollows in the bedrock. The potholes now visible in the bottom of the canyon are merely modern examples of a continuous succession of potholes that are almost wholly responsible for cutting the canyon. This mechanism is significantly diffferent from that of Enfield Glen,

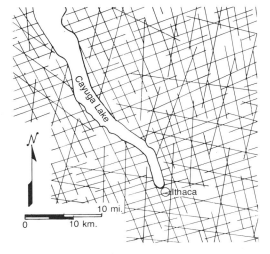

Joint (fracture) patterns of the Cayuga Lake region. These control the erosion in Enfield and Taughannock glens.
Adapted from NYSGA Guidebook, 31st annual meeting, Cornell University, 1959.
Courtesy Kendall / Hunt Publishing Co.

where the stream follows joints and, during flood, lifts and transports joint-bounded slabs downstream. Lack of joint control seemingly gave the postglacial Glen Creek greater freedom to seek a new course independent of the pre-Wisconsin one. As a result, the old, drift-filled gorge is intersected only at one point, near the head of the modern gorge.

Other glens nearby are at Shequaga Creek, Johns Creek, Catlin Creek, and Havana Glen. Others farther south are actually cut partly into the Valley Heads moraine, as at Sleeper Creek and Johnson Hollow.

The road hugs the west wall of the trough between Watkins Glen and Montour Falls (2 miles), passing almost continuous roadcuts in Genesee group shales and siltstones. This section of valley floor is marshy lake plain referred to as Bad Indian Swamp.

Between Montour Falls and Horseheads (14 miles), you follow Catherine Valley through, rather than over, a vast expanse of Valley Heads moraine. Catherine Creek is the northflowing feeder to Seneca Lake that has breached the divide, in places cutting a 200-foot deep canyon in the till, and causing intricate, badlands-type dissection along many small side streams. Glacial till is ideal for badlands erosion because it is unconsolidated, more or less uniform in its response to erosion, and capable of standing in steep slopes. Here, the dissected slopes are largely concealed by vegetation, but a magnificent display of barren badlands may be seen at Chimney Bluff on the Ontario shore just east of Sodus Bay (see US 104: Rochester—Oswego discussion). The moraine grades quickly into pitted outwash plain about 2 miles north of Horseheads. For a discussion of the numerous glacial features around Horseheads, see NY 17: Binghamton—Arkport/Hornell.

Watkins Glen, Rainbow Falls, cut in shales and sandstones of the Enfield formation (Devonian)

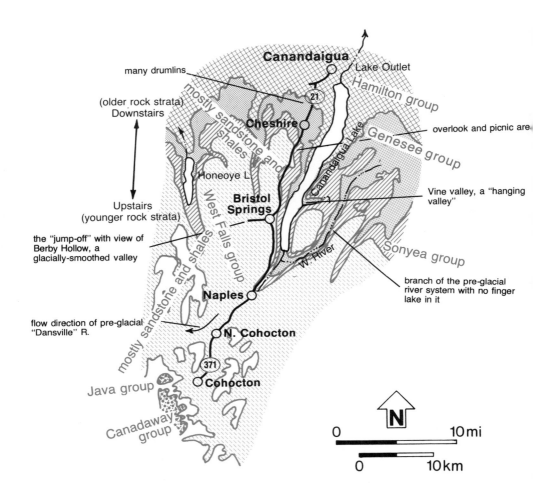

Canandaigua

Lake Outlet

many drumlins

Hamilton group

(older rock strata)
Downstairs

overlook and picnic area

mostly sandstone and shales

Cheshire

Genesee group

Honeoye L.

Upstairs
(younger rock strata)

West Falls group

Bristol
Springs

Canandaigua Lake

Vine valley, a "hanging valley"

the "jump-off" with view of
Berby Hollow, a
glacially-smoothed valley

W. River

Sonyea group

branch of the pre-glacial
river system with no finger
lake in it

flow direction of pre-glacial
"Dansville" R.

mostly sandstone and shales

Naples

N. Cohocton

371

Cohocton

Java group

Canadaway
group

N

| 0 | 10 mi |
| 0 | 10 km |

**NY 21, NY 371
CANANDAIGUA—COHOCTON**

NY 21, NY 371:
Canandaigua—Cohocton
37 mi./50 km.

This is one of the best routes on which to witness the true spectacle of the Finger Lakes. The route stays high on the west wall of the Canandaigua Lake trough in the northern 15 miles, with numerous open views over the Allegheny Plateau and the lake far below. The lake occupies the main trunk valley of a branching preglacial stream system. The dendritic, or tree-like, drainage pattern, points southward, indicating that the preglacial drainage was toward the south. This is true, in fact, in all of the Finger Lakes West, i.e., all lakes west of Seneca Lake. The principal valleys that trend generally north-south have been deepened and widened by ice erosion into broad, steep-sided troughs, while the divides and upland hills have been streamlined by the overriding glacier. The net result is a uniquely beautiful landscape of very large scale. At the highest point on NY 21, the road is about 700 feet above the lake, but still well below the height of the upland surface, and nearly 6 miles from the eastern divide across the lake. In surveying this lovely scene from a quiet place along the road, the impressions of height, breadth, and natural splendor are akin to those of the Grand Canyon of the Colorado River, although the smooth, subtler outlines are inherently more tranquil. A truly distinctive aspect of this and all other glacial troughs is the way the walls "roll" downward and steepen toward the lake. Looking out over this plunging slope to the lake, the near shore is often hidden from view, and the land seems suspended high above the water—and the spirit soars! All of these impressions are enhanced, of course, by

Hemlock (left) and Canadice lakes from Straight Road on hill near surface. Courtesy Kendall/Hunt Publishing Co.

knowledge of the momentous geologic events that produced this scenery, and by the ability to read history in the tell-tale language of the landscape. That, in fact, is what this book is all about!

Just southward of Canandaigua, the road passes dozens of drumlins. The drumlins may be so numerous here simply because the surface rises to the south on the approach to the Portage scarp (see US 20: Buffalo—Canandaigua discussion). About a mile south of Cheshire, two small roadcuts expose dark shales that are stratigraphically near the contact between the Devonian Genesee (lower) and Sonyea (upper) groups. Two and one half miles south of Cheshire, there is a small picnic area by the road with incredible views of the northern half of the lake and surrounding landscape. Note the persistent rise to the south of the upland surface viewed across the lake and the accordance, or alignment, of the hilltops. Similar features are present in nearly all of the Finger Lakes. On the opposite side of the lake a little south of this spot, the even profile is broken by Vine Valley, the northernmost of the major tributary troughs that comes in from the northeast. Scissored at the juncture is Bare Hill, with its distinctive triangular profile of long, tapered, north, or "stoss," slope smoothed by southward-moving ice, and steeper, south slope truncated by the Vine Valley ice tongue. Vine Valley is really a "hanging" valley, with a bedrock floor well above that of the Canandaigua

234

Springwater; note higher elevation of Canadice, and accordance of upland

trough. Such valleys are abundant in the Finger Lakes region. They are formed where preglacial tributary streams joined the main valleys at angles less suited to deepening by ice, i.e., more athwart the ice flow. When the ice vacated the land, these valleys were left high and dry on the sidewalls of the main troughs. Deep incision by modern streams has produced the many waterfalls and glens, or gulfs, hollows, or gullies for which this region is famous. The nearly level floor of Vine Valley is only about 150 feet above the lake, but the maximum depth of the lake is about 275 feet, and the bedrock floor is an untold great depth below that. Therefore, Vine Valley must "hang" considerably more than 425 feet above the bedrock floor of the Canandaigua trough.

With good visibility from the picnic area, you can just barely see over the eastern divide and into the West River trough, which joins the Canandaigua trough at the south end of the lake.

About a mile south of the picnic area, at the crest of a hill, the southern end of the lake is visible with its extensive delta built by the West River. The southern ends of most of the Finger Lakes are much deeper and fjord-like than the northern ends because of the higher elevation of the adjacent upland surface. In addition to the extensive river delta at the southern end, there are numerous, small, fan-shaped deltas along the lakeshore built by tiny, inflowing streams as they cut deep gashes in the sidewalls. The largest of these is Seneca Point, not far south of here on the western shore.

One of the best places to see a glacial trough in profile is near Bristol Springs in Ontario County Park. The park is perched on top of Gannett Hill, 1.5 miles west of Bristol Springs; and the viewpoint is an overlook called "The Jumpoff," located on the western border of the park at the lip of the West Hollow trough. The profile to see is not of West Hollow itself, but of its northern extension, called Berby Hollow, that begins about a mile north of the Jumpoff to the right. The cross-section is incredibly smooth and U-shaped, distinctively the product of glacial erosion. A similar bedrock profile may be depicted in the lake troughs, but deeper. The curvature has been described as "catenary," meaning it is like that of a link chain suspended from two fixed points. Its smoothness in Berby Hollow suggests that there is little or no glacial till on the valley floor. By contrast, West Hollow, below and left of the Jumpoff, has a moderately mounded floor resulting from morainal deposition.

From Bristol Springs south, NY 21 descends the west wall to the shore, past three small creek deltas at Walton, Grange, and Coy

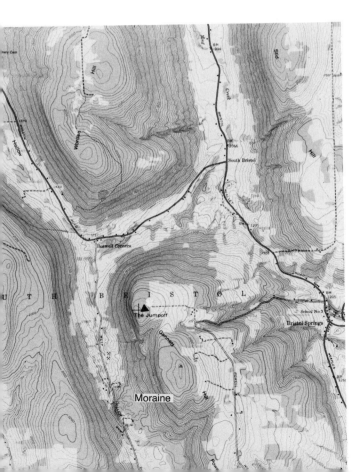

Section of the Bristol Springs 7.5-minute topographic quadrangle map showing Berby Hollow and the branching glacial trough of Mud Creek. The smooth U-shaped surface clearly evident in the topographic contours is broken by mounded moraine south of The Jumpoff. Courtesy Kendall/Hunt Publishing Co.

Smooth-curve profile of Berby Hollow, a glacially scoured valley near Bristol Springs

points, and then skirts the large, valley-filling delta of West River to Naples past High Tor Game Management Area and Parish Flat. Most of the way, South Hill looms large and imposing across the valley, and then terminates abruptly at the flat sediment plain where West River trough comes in. The river itself comes down the trough from the northeast, then makes a sharp turn to north around the South Hill buttress, called Sunnyside Point, and empties into the lake.

Naples is a village steeped in history. Now one of the most important winery centers in New York, it is the site of the original Seneca Indian village of Nundawao, which was later renamed Watkinsville, then Middletown, and finally, in 1808, Naples. It is the burial place of Conesque, Chief of the Senecas, who came "home" from the Genesee Reservation in 1794 to die here. This is also the place where General George Washington first took command of the American Army on July 3, 1775. A plaque in the village park marks the spot where he stood on this momentous occasion that was to profoundly influence the course of American history.

Naples is right at the place where the delta plain merges with the Valley Heads moraine, at an elevation of about 800 feet, and just north of Italy Valley, another tributary trough that comes in from the east. The topography changes dramatically south of Naples on NY 21 to one of hummocks and hollows as the surface climbs steeply to nearly 1400 feet at North Cohocton (5 miles). In typical fashion, once the moraine is topped, a mile or so before North Cohocton, the hilliness quickly diminishes at the beginning of the pitted outwash plain. Kettles here are all rather small and shallow, but numerous.

NY 21 turns westward at North Cohocton along a cross-trough to Wayland (8 miles). This is the channel of the preglacial ancestor of the Genesee River, the so-called "Dansville River." Between North Cohocton and Cohocton along NY 371 (5 miles), the route follows the level, locally marshy, outwash plain along the Cohocton River. The sides of the valley en route are extensively terraced with gravelly, sandy sediments indicative of either outwash plain stream deposits, or marginal ice deposits called kames. The terraces stand 50-100 feet above the river.

Wind-etched glacial outwash sand

Interstate 390:
Rochester—Avoca
70 mi./113 km.

This route cuts across the east-west bands of Silurian strata that floor the Ontario Lowlands and Devonian beds of the Catskill Delta that underlie the Allegheny scarp and plateau. All dip gently southward, and because of this and long-term downwasting of the surface, the bands are arranged in order of decreasing age from north to south. A traverse in that direction "climbs the section" and offers a progressive account of the course of geologic history. Numerous north-south routes in the Finger Lakes region do this, but this one probably has the most roadcuts. Keep in mind, however, that it is illegal to stop on the Interstate highway.

The scarp slope of the Allegheny Plateau in this region appears to have been ground down by the direct onslaught of ice that moved southward out of the Ontario basin. Elsewhere, where the ice front advanced diagonally against the scarp, it was less severely reduced, and steep slopes or cliffs remain today. The lowlands/plateau boundary is placed just south of Rush at the northern limit of the Devonian strata; from there southward, the land rises gradually.

At about 6 miles south of Rochester, near the Interstate 90 junction, there is a subtle, 100-foot high scarp from which you have a sweeping view to the north of the city of Rochester. Visible are the Rochester plain, the drumlins, and the Pinnacle Hills moraine-kame complex of south Rochester with the radio tower on it, a part of the Rochester-Albion moraine belt (see US 104: Lewiston—Rochester discussion). The Mendon Ponds moraine-kame-kettle complex, rep-

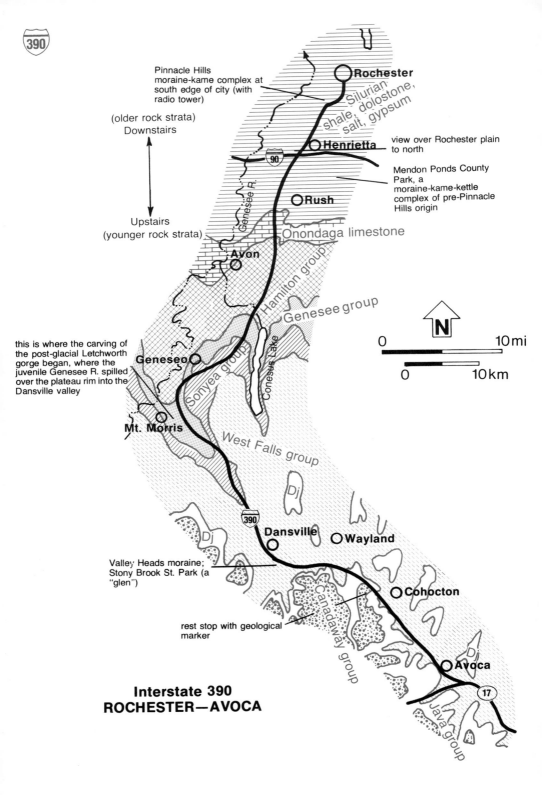

390

Pinnacle Hills
moraine-kame complex at
south edge of city (with
radio tower)

Rochester

Silurian
shale, dolostone,
salt, gypsum

(older rock strata)
Downstairs

view over Rochester plain
to north

Henrietta

90

Mendon Ponds County
Park, a
moraine-kame-kettle
complex of pre-Pinnacle
Hills origin

Genesee R.

Upstairs
(younger rock strata)

Rush

Onondaga limestone

Avon

Hamilton group

Genesee group

this is where the carving of
the post-glacial Letchworth
gorge began, where the
juvenile Genesee R. spilled
over the plateau rim into the
Dansville valley

Conesus Lake

Sonyea group

Geneseo

Mt. Morris

West Falls group

Dj

390

Dansville

Wayland

Valley Heads moraine;
Stony Brook St. Park (a
"glen")

Dj

Cohocton

rest stop with geological
marker

Canadaway group

Avoca

Dj

17

Java group

**Interstate 390
ROCHESTER—AVOCA**

resenting a stand of the receding Wisconsin ice earlier than the Pinnacle Hills, lies about 5 miles east of this point. During Pinnacle Hills time, glacial Lake Dana filled these lowlands, and some of the higher kame hills at Mendon Ponds stood up as islands.

The road crosses Honeoye Creek 4 miles south of Interstate 90. The creek drains Hemlock, Canadice, and Honeoye finger lakes, located directly to the south. At this point it occupies a portion of the preglacial Genesee Valley and flows westward a few miles to join the modern Genesee. A few miles to the east, the creek comes in from the south and is separated by a low drainage divide from the headwaters of tiny Irondequoit Creek which continues down the old Genesee Valley to Irondequoit Bay, the drowned mouth of the former river (see US 104: Rochester—Oswego discussion).

The denser part of the drumlin field ends near Avon, 6 miles south of Honeoye Creek. The most striking geologic features between Avon and Dansville (30 miles) are related to the Genesee River, past and present. The highway converges with the Genesee Valley at Geneseo and follows it for 2 miles to Mt. Morris. This is a section of the river that continues northward to the Honeoye Creek junction, where it re-occupied the master stream valley of the pre-Wisconsin drainage system. Hence, this section is much wider than the Letchworth or Rochester gorges where solid bedrock had to be excavated. The Genesee River in this part meanders widely across a broad, flat, sediment-filled floor; "ox-bow" lakes are common. Oxbows form when shortcutting floodwaters abandon meander loops. The river is said to be underfit, to small to have carved its large valley.

Mt. Morris is where the Letchworth Gorge started, where the juvenile Genesee spilled over the rim of the Letchworth plateau into the Dansville trough in the waning stages of Wisconsin glaciation. As is the case at Niagara, the headward recession of the falls carved the canyon. The highway leaves the Genesee at Mt. Morris and follows the Dansville trough, now occupied by miniscule Canaseraga Creek, for 15 miles to Dansville. The so-called "Dansville River" was the east branch of the ancestral Genesee, with headwaters at the site of the present Canandaigua Lake. From there, the river flowed south-southwestward to North Cohocton, then west to Dansville, then northwest down the Dansville trough. The west branch, meanwhile, originated in Pennsylvania near the New York border and flowed generally northward, joining the east branch at Sonyea, located between Mt. Morris and Dansville. Note the broad, deep, west branch valley as you go by Sonyea, a valley now occupied by tiny Keshequa Creek. A few roadcuts nearby expose shales of the Genesee and Sonyea groups. A segment of Valley Heads moraine near Portageville at the south end of Letchworth Gorge prevented the modern river from reaching the old channel of the west branch. Another segment of moraine plugs the old eastern channel near Dansville.

This is lovely, open country of high relief with sweeping views. One cannot help but be impressed with the immensity of the old valleys and the enormous volume of water they must have carried. At Dansville, you pass Stony Brook State Park, at the bedrock gorge of a small brook incised into the wall of the trough. Canyons like this are a distinctive trademark of the Finger Lakes region, and most are referred to as "glens." All are carved by post-glacial streams that spilled off the uplands into the deeply-gouged glacial troughs, forming falls. Many were initiated in the approximate routes of pre-Wisconsin glens, and are now cut partly into bedrock and partly into old channels that were filled with glacial drift. In many respects, the Stony Brook Glen resembles the Enfield Glen south of Ithaca. The rocks exposed are all late Devonian, mostly Nunda formation sandstones and shales belonging to the West Falls group, essentially the same rocks that support the upper and middle falls in the Letchworth Gorge.

Between Dansville and Wayland (6 miles), the road climbs 700 feet east over the Valley Heads moraine, to the head of east-west-trending North Cohocton-Dansville Valley. The moraine blockage constitutes a through-valley divide separating Canaseraga Creek from the opposite drainage of the Cohocton River. The upper valley floor near Wayland is pitted outwash with several small kettle lakes.

Between Wayland and Cohocton (6 miles), the road goes overland to the Cohocton Valley with several, excellent, new roadcuts in interbedded shales, siltstones, and sandstones of the late Devonian Java and Canadaway groups.

The Cohocton River is one of the major tributaries of the Susquehanna River, which it joins at Corning (see NY 17: Binghamton—Arkport/Hornell discussion). The valley is also a typical glacial trough with all of the tell-tale characteristics, such as a broad, flat, sediment-filled floor, abrupt steep walls, truncated ridge endings, or spurs, hanging side valleys, and an underfit river. An excellent rest stop and overlook near Cohocton has a geological marker explaining some of these features.

INTRODUCTION TO LETCHWORTH GORGE

Letchworth Gorge, the "Grand Canyon of the east," is the 22-mile section of the Genesee River between Portageville and Mt. Morris that is now under state protection as Letchworth State Park. It is a portion of the river whose meandering course has been deeply incised into the plateau since the end of the Ice Age, producing remarkably narrow, winding bedrock canyons at either end with nearly vertical walls. The upper canyon at the Portageville end contains three major cataracts, and the Mt. Morris Highbanks canyon at the lower end is the site of the Mt. Morris flood-control dam. In the central section, the river now occupies a preglacial valley with high, but gently-sloping sides; it meanders over a moderately broad flood plain.

SITES OF GEOLOGIC AND SCENIC INTEREST IN LETCHWORTH GORGE

1. Upper Falls
The caprock of the upper falls is resistant sandstone of the Nunda formation. This is supported by the underlying sandstone and shale of the Gardeau formation, beginning about 28 feet below the rim. Both units belong to the late Devonian West Falls group of the Catskill Delta.

In this section of the gorge there are three cascades in 2 miles, the upper, middle, and lower falls over which the river descends 71, 107, and 70 feet respectively. Since all three receive the same discharge they have most likely receded synchronously, leaving the gorge in their wakes. But why are there three falls instead of one? The answer to this question probably lies in the formation of temporary meltwater lakes downstream that controlled the baselevel of erosion for each falls. The upper falls formed when the river spilled over a scarp bordering the highest lake. When they had retreated for some distance the lake level dropped in response to the opening of a new, lower outlet; and the middle falls were initiated at the scarp where the upper falls had been in the first place. Both upper and middle falls then receded at essentially the same rate until the lake level dropped again and the lower falls formed. The controlling lake possibly occupied the broad valley in the central section of the park.

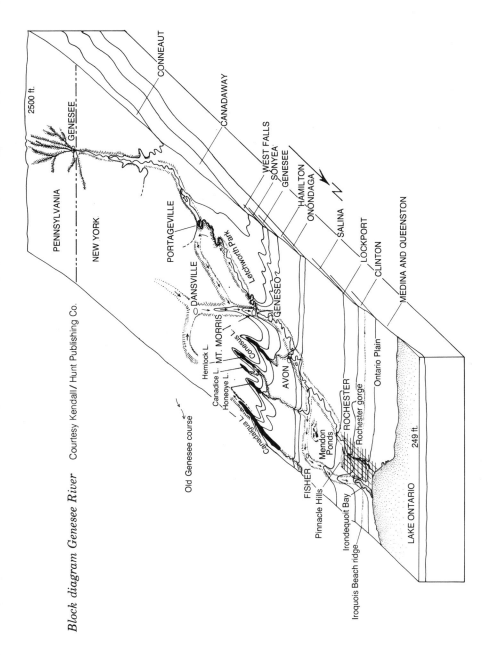

Block diagram Genesee River   Courtesy Kendall/Hunt Publishing Co.

The impressive Portage railroad bridge behind the upper falls is 234 feet above the river. The steel structure replaces a handsome wooden bridge that was built in 1852 and destroyed by fire in 1875. At the time it was the largest, all-wooden bridge in the world, requiring timber from 300 acres of woodland. The overgrown bed of the abandoned Genesee Valley Canal is notched into the east wall of the gorge under the railroad bridge. The canal provided a connecting link between the Erie Canal at Rochester and the Allegheny River at Olean, but it was commercially unsuccessful. It came up the Dansville Valley to Sonyea and then rose about 520 feet through 50 locks to the Portageville level near the village of Oakland, passing part of the way along the gorge wall. The canal was opened from Rochester to Portageville in 1851 and to the Allegheny River in 1857. It was completely abandoned in 1882 and parts of it were used as railroad bed between 1880 and 1963. It was the railroads more than anything else that led to the demise of the canal systems because they offered cheaper and faster transportation of goods. Adding insult to injury many of the newer railroad lines used the level, abandoned stretches of the old canals.

2. Middle Falls

The middle falls, located a half mile downstream from the upper falls, are the most scenic of the three falls. They are capped by resistant sandstone beds within the Gardeau formation. One of the main reasons for believing that the three falls originated at three different glacial lake levels at different times is that the rock section is more or less uniform in its resistance to erosion. There are no distinct caprock beds like the Lockport dolostone of Niagara Falls, so it is unlikely that an original single falls would have separated into three falls as it receded.

A landslide area on the east wall of the canyon about 500 feet downstream from the falls is marked by a fan-shaped pile of debris at river level. The sliding occurs in unconsolidated glacial till directly above the fan that fills a pre-Wisconsin river valley cross-cut by the modern river. The bed of the Genesee Valley Canal passed through the slide area and was repeatedly damaged by slippage. North of the slide, the canal builders attempted to blast a tunnel through bedrock to short-cut the next bend in the river where the gorge wall is nearly vertical. However, cave-ins soon occurred and the effort had to be abandoned.

Upper falls in Letchworth State Park, from railroad bridge; abandoned canal just to right   Courtesy Kendall / Hunt Publishing Co.

3. Lower Falls

A pathway leads down into the canyon to the lower falls and crosses a footbridge over the Flume, a narrow neck of the gorge. A pinnacle on the east side called Cathedral Rock became an island during the high water caused by Hurricane Agnes in 1972. The water level then was just 4 feet below the top of the Mt. Morris flood control dam 17 miles downstream. At the Flume, a large flat surface of resistant sandstone called Table Rock, was swept clean by sidesweeping of the river before the Flume was excavated. Large-scale ripple marks on the surface probably formed by tidal currents when the original sands were deposited in a shallow sea.

4. Rim Overlooks

Several rim overlooks provide spectacular views of the gorge. The Big Bend overlooks are highest, with 550-foot walls. The course of the river is so tortuous that it is difficult to know what section of canyon you're seeing at any one point, or which way the river should be flowing. Rivers don't normally wind much when they are so confined by rock walls. This one does because its winding course was first established on the plateau where it had freedom to move sideways. Falls regression along the sinuous course left a canyon of similar form.

5. Gardeau Overlook

From the parking area for this overlook, a path leads down to the floor of the broad, central valley of the park where the river is

Winding Letchworth Gorge, in sandstones and shales of the West Falls group (Devonian)

unconfined and meanders freely across the wide floodplain. The flats of this section were, for 50 years, the home of Mary Jamison, the "White Woman of the Genesee," whose remains are now buried near the middle falls. For most of her life, she was a captive and adopted member of the Seneca tribe.

6. Mt. Morris Highbanks Gorge

Several rim overlooks in this section include those by the Mt. Morris Dam. The dam, which was completed in 1952, was designed to reduce the threat of flooding in the lower Genesee Valley. The record shows that severe floods plagued the valley as far back as 1800. Terrible destruction resulted from flooding in 1865 when discharge reached half the volume of the Niagara River. A major flood ravaged the region on the average of once every seven years from 1865 to 1950.

There are excellent views of the Hogsback from the Highbanks Recreation Area. The Hogsback is a knife-edged meander divide worn almost completely away by sidesweeping of the river in a giant meander loop. An oxbow lake may also be seen there in the floodplain, but the configuration may change from year to year as a result of filling of the reservoir.

7. Da-Yo-It-Ga-O

This overlook near the north entrance to the park offers a picturesque view into the Dansville-Mt. Morris trough.

8. Other Scenic Attractions of the Letchworth Area

The Mt. Morris Dam may also be approached from the other side of the river by way of an entry road that leaves NY 408 one mile west of Mt. Morris. The Mt. Morris Highbanks gorge was carved by recession of falls that spilled over the wall of the Dansville-Mt. Morris trough by the village of Mt. Morris. The southern end of NY 408 descends to Nunda, with excellent views into the Nunda-Sonyea Valley.

Zoar valley, carved in Devonian Canadaway shale

US 62, NY 241, NY 17:
Hamburg—Steamburg
47 mi./75 km.

Hamburg lies on the gentle slopes of the Erie plain at the base of the Allegheny scarp, here called the Erie scarp. It is also at the mouth of a large glacial trough furrowed into the scarp to the southeast, and now occupied by tiny Eighteenmile Creek. The scarp has been similarly furrowed in several other places along the Erie front by ice that advanced generally southeastward from the Lake Erie basin, pushing tongues upstream in preglacial river valleys and gouging them out (see maps with NY 17: Arkport/Hamell—Westfield).

Between Hamburg and Eden (5 miles), US 62 crosses the north and south branches of Eighteenmile Creek which expose shales of the late Devonian West Falls group. Many small creeks like these have incised the scarp and lower slopes since the close of the Ice Age. Eden is at the foot of a 100-foot-high ledge of the overlying Java group sandstone and shale. A short distance upslope is a steep slope developed on the overlying Canadaway group shales and siltstones.

Between Eden and Gowanda (15 miles) is flat, open farmland bordering a gentle slope east of the highway, all underlain by Canadaway bedrock. Many old gas wells around Gowanda have produced from the Silurian Medina sandstone. Gowanda occupies one of three key points of the most far-reaching, glacially-influenced drainage changes of the Allegheny Plateau. Cattaraugus Creek passes through the village where it exits the deep gorge of the postglacial Zoar Valley and enters onto the broad, open, northwestern reach of the preglacial valley that was part of the ancestral Allegheny River system.

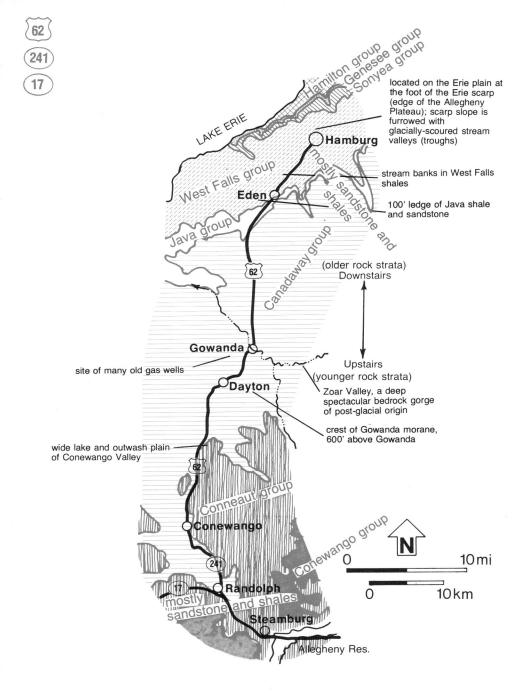

located on the Erie plain at the foot of the Erie scarp (edge of the Allegheny Plateau); scarp slope is furrowed with glacially-scoured stream valleys (troughs)

stream banks in West Falls shales

100' ledge of Java shale and sandstone

(older rock strata) Downstairs

Upstairs (younger rock strata)

Zoar Valley, a deep spectacular bedrock gorge of post-glacial origin

crest of Gowanda morane, 600' above Gowanda

site of many old gas wells

wide lake and outwash plain of Conewango Valley

Hamilton group

Genesee group

Sonyea group

LAKE ERIE

Hamburg

West Falls group

Eden

Java group

mostly sandstone and shales

Canadaway group

Gowanda

Dayton

Conneaut group

Conewango

Conewango group

N

0   10 mi

0   10 km

Randolph

mostly sandstone and shales

Steamburg

Allegheny Res.

NY 62, NY 241, NY 17
HAMBURG—STEAMBURG

The maps on page 172 compare the preglacial drainage pattern with the present one. Presently, the Allegheny River drains southward to the Ohio River and then to the Mississippi River. The preglacial river flowed northward to Gowanda, northwestward to the "Erian River," and then on to the Atlantic via the St. Lawrence Valley.

At Gowanda, the re-established postglacial drainage, now Cattaraugus Creek, was prevented by our old friend, the Valley Heads moraine, from extending southward to the preglacial valley. Instead it cut away at the bedrock to the east and carved the Zoar Valley. The gorge is cut 300-400 feet deep into the shales and siltstones of the Canadaway group. Because the canyon is youthful, geologically speaking, and the weak beds are stacked like books, the walls are nearly vertical.

Other key points that influenced the changes in the Allegheny system are discussed below. Between Gowanda and Dayton (5 miles), the road climbs nearly 600 feet to the crest of the moraine via a narrow canyon. This north slope of the moraine is deeply and intricately dissected by innumerable small rivulets that are attempting to reduce the landscape to the Cattaraugus Creek baselevel. Presently their headwaters are clawing into the outwash plain south of the moraine.

The route follows the outwash/glacial lake plain between Dayton and Steamburg (30 miles) through the wide, shallow Conewango Valley. Its upper end near Dayton is an extremely broad, marshy plain pockmarked with tiny kettle lakes and scratched by a chaos of disjointed rivulets.

En route to Randolph, the road leaves the valley to pass briefly through the bordering hills of late Devonian Conneaut group shales and siltstones, revealed in a few roadcuts.

Conewango Creek is a grossly underfit stream that wanders lazily over the valley flats like a snake, leaving multiple meander scars. A few miles south of Conewango, it turns sharply west and then south to join the Allegheny at Warren, Pennsylvania. The preglacial Allegheny flowed in the opposite direction, as shown by the maps. This section of the valley held the main trunk stream. Obviously it must have been a very large river.

Between Conewango and Randolph (8 miles), you follow NY 241 along the valley side. Then, from Randolph to Steamburg (8 miles), you follow NY 17 on the opposite side. Little Conewango Creek flows northwestward through this segment of the preglacial valley in the same apparent direction as that of the ancestral Allegheny.

Steamburg is at the border of the so-called Salamanca re-entrant, the only area of New York that escaped glaciation during all four major advances ot the Ice Age. The re-entrant is a roughly triangular area enclosed by the Pennsylvanian border and the course of the Allegheny River in its short traverse into New York and back to Pennsylvania. At the climax of Wisconsin glaciation, the ice front stood just north of the present course of the river, blocking the Conewango drainage route and forcing an escape to the south. In time, the new southern channel was sufficiently established so that the river continued to use it even after the ice dam had melted away. The valley at Steamburg is now marked by a low end moraine that is largely buried in outwash and modern alluvial fan and deltaic sediments. This divide helped assure that the postglacial Allegheny would continue in its southerly route. Thus, Steamburg also occupies a strategic position in the evolution of the Allegheny system.

The most important element that led to the current Allegheny drainage system was the erosional reduction of a preglacial divide in Pennsylvania, called the Kinzua col, where the Kinzua Dam is now located. The preglacial drainage was northward from the Kinzua col to Steamburg; now it is southward. The col was worn down by spiller of the copious meltwater that filled the Allegheny and branching valleys, in the section of valley that now contains the Allegheny Reservoir. By the time Wisconsin ice retreated, the divide was obliterated, and the southerly drainage route was well entrenched.

Tully Lakes, a fine group of kettle lakes, as seen from the Song Mountain ski slopes. I-81 is in the mid-distance on the right at nearly the same level as the moraine; the moraine begins a drop of 600 ft. to the trough floor just beyond the lakes.

Interstate 81:
Syracuse—Pennsylvania State Line
80 mi./130 km.

Syracuse is at the northern margin of the Allegheny Plateau, where it joins the Ontario Lowlands. South of the city on Interstate 81, the most striking geologic features are of glacial origin: glacial troughs, moraine, outwash plains, and kettle lakes. In fact, this landscape characterizes the entire Finger Lakes portion of the Allegheny Plateau. During the Ice Age, when glaciers invaded from the north, certain valleys in line with the ice flow were gouged deeper and wider. Later, during deglaciation, when they were the sites of melt-water lakes, these troughs received thick deposits of till, outwash sediments, lake clays, and delta sediments. The net result is that the original stream valleys are now very broad, with steep sides and flat floors. A prominent feature in many of them is the Valley Heads moraine, which divides the north-flowing and south-flowing drainage of the valley floor. The moraine was formed during a long pause, perhaps 12,000-14,000 years ago, in the northward recession of the Wisconsin ice sheet, when great mounds of glacial debris were dumped in the troughs along the ice margin. Segments of the Valley Heads moraine may be found damming the southern ends of all of the Finger Lakes. One segment, called the Tully moraine, is visible from Interstate 81, 12 miles south of Syracuse, but here there is no Finger Lake north of it.

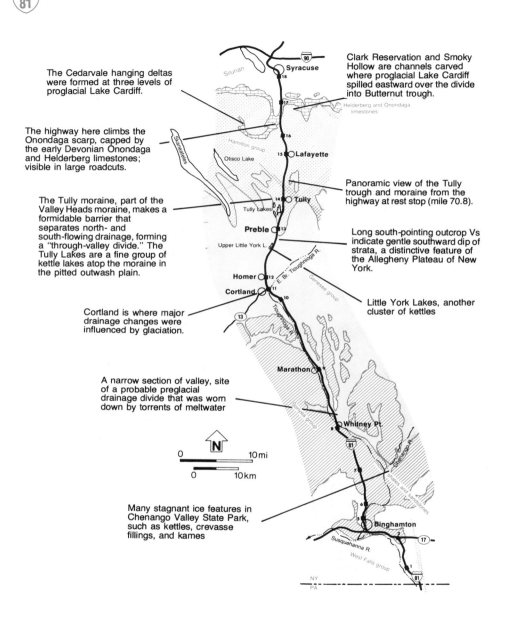

The Cedarvale hanging deltas were formed at three levels of proglacial Lake Cardiff.

Clark Reservation and Smoky Hollow are channels carved where proglacial Lake Cardiff spilled eastward over the divide into Butternut trough.

The highway here climbs the Onondaga scarp, capped by the early Devonian Onondaga and Helderberg limestones; visible in large roadcuts.

Panoramic view of the Tully trough and moraine from the highway at rest stop (mile 70.8).

The Tully moraine, part of the Valley Heads moraine, makes a formidable barrier that separates north- and south-flowing drainage, forming a "through-valley divide." The Tully Lakes are a fine group of kettle lakes atop the moraine in the pitted outwash plain.

Long south-pointing outcrop Vs indicate gentle southward dip of strata, a distinctive feature of the Allegheny Plateau of New York.

Little York Lakes, another cluster of kettles

Cortland is where major drainage changes were influenced by glaciation.

A narrow section of valley, site of a probable preglacial drainage divide that was worn down by torrents of meltwater

Many stagnant ice features in Chenango Valley State Park, such as kettles, crevasse fillings, and kames

Interstate 81
SYRACUSE—PENNSYLVANIA STATE LINE

254

Onondaga Lake, which lies partly in Syracuse, is a Finger Lake of a different sort, since it is not nestled in a deep, steep-sided valley like those in the west; in fact, it is completely within the Ontario Lowlands. The southward extension of the lake depression into the uplands is the Onondaga trough, which continues as the Tully Valley after its junction with the west branch Onondaga trough near the village of South Onondaga. South of the city, the road emerges from the Onondaga trough past several excellent roadcuts in the Helderberg and Onondaga limestones. The limestone beds are slightly folded into broad shallow anticlines or upfolds, and synclines or downfolds. Folding like this is common in the upland country; the severity of folding increases southward toward the Appalachian Mountains of Pennsylvania (see Genesee Valley discussion).

Views to the north over Syracuse, Onondaga Lake, and the Ontario plain are especially good between miles 77 and 75, and they are spectacular at night. At mile 71.5, you cross a deep ravine called Webb Hollow which, near its mouth at the Onondaga trough, is 900 feet deep. Clefts like this, which are also called "gulfs" and "glens," are a trademark of the Finger Lakes region, whose origin is inextricably tied to that of the glacial troughs. During the ice advance, when the troughs were being deeply gouged, the lesser cross valleys were simply passed over, and many were filled up with glacial debris. After the ice receded northward, new tributary streams, initiated on the upland surface, cascaded over the high walls of the troughs. The canyons have since formed in the wake of the falls as they slowly migrated upstream, much as the Niagara Gorge was carved over the last 12,000 years. Several such gulfs are visible on the west wall of the Tully Valley from Interstate 81 farther south. Rattlesnake Gulf, for example, is visible directly across from the rest area at mile 70.8, where the view of the valley itself is breathtaking.

Views of Tully Valley and the Tully moraine to the south are especially good between miles 71 and 69. The moraine rises more than 500 feet above the valley floor and levels off at about road level near Exit 14 to Tully, where a broad cross valley comes in from the east. Contemporaneously-formed segments of the Valley Heads moraine are found in the Otisco Valley trough to west, called Bennett Hollow, and the Butternut trough to east, called Clark Hollow. These moraine blockages are called "through-valley divides" in reference to their function as drainage divides within a continuous bedrock valley.

The Tully Lakes, visible atop the moraine from near the Tully exit, are a cluster of kettle lakes. Such lakes are common in the "pitted outwash plains" immediately downvalley from the Valley Heads moraine, or any other moraine, for that matter. They form where blocks of ice calve from the wasting ice front and become partly or entirely buried in outwash sediments before they melt totally. The resulting kettle holes may or may not have water in them. Kettle lakes may often be recognized by their steep banks of outwash materials, rounded outline, and poorly integrated drainage, both inflowing and outflowing. The Tully Lakes may be viewed to best advantage from the top of Song Mountain just south of them. The lakes are at the headwaters of the west branch Tioughnioga River, which joins the east branch at Cortland, 15 miles to the south.

The Otisco trough, the southward extension of Otisco Lake, easternmost of the major Finger Lakes, joins the Tioughnioga trough at Preble in a Y-junction that points sharply southward. The valley floor here is exceedingly broad and flat, with abrupt steep walls, suggesting very deep sediment fill in its floor. A mile south of Preble, the highway passes close by the west wall of the valley next to the Little York lakes, another cluster of kettles. A similar, but smaller, Y-junction occurs at Homer, where the southern extension of the Skaneateles Lake trough, or Factory Brook valley, joins the main valley.

Cortland obviously lies at the apex of important drainage changes caused by Wisconsin glaciation. The city is built on a broad outwash plain at the juncture of several main valleys, including the west and east branches of the Tioughnioga River, Trout Brook from the east, the broad valley occupied by tiny Otter Creek from the southwest, and the main branch Tioughnioga River extending to southeast. The latter valley differs markedly from the others in being almost V-shaped with only a very narrow floodplain, indicating a somewhat different origin. The thick sediment fill at Cortland actually extends 4-5 miles up Otter Creek valley as pitted moraine and outwash, and

Looking south from Clark Hollow into glacial trough of Labrador Pond

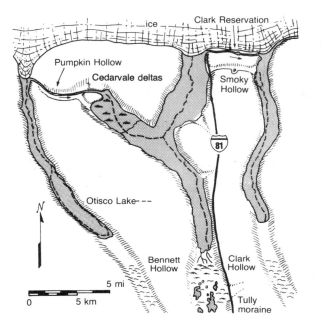

*Proglacial Lake
Cardiff (center)
during Wisconsin
deglaciation,
showing
divide-breaching
spillover channels at
Pumpkin Hollow,
Smoky Hollow, and
Clark Reservation,
and 3 segments of the
Valley Heads
moraine in the south*
Courtesy Kendall/ Hunt
Publishing Co.

constitutes a through-valley divide separating the southwest-flowing
Fall Creek from the east branch Tioughnioga River. The divide is
another segment of the Valley Heads moraine. The preglacial Fall
Creek had its origin at the head of what is now the east branch
Tioughnioga near Labrador Pond, Fabius, and Sheds, and flowed
straight southwestward through the Cortland area to the Cayuga
River, whose widened and deepened valley is now the Cayuga Lake
trough.

Three miles north of Exit 11 to Cortland, you have an exceptional
open view of the Cortland plain. Several large cuts near this exit are
in dark Devonian shales that typify much of the Allegheny Plateau
bedrock. Between Cortland and Whitney Point the road follows the
narrow valley of the main branch Tioughnioga River to its junction
with the Otselic River. The narrowness of this valley, as compared to
the broad glacial troughs elsewhere, requires explanation. In the first
place, this valley appears to have been much smaller before it was
gouged out by the advancing ice. Secondly, and even more important,
there were really two aligned streams, one flowing northward to the
preglacial Fall Creek, and the other southward from a bedrock divide
that existed about halfway between Blodgett Mills (4 miles from
Cortland) and Messengerville (10 miles). The southward advance of
the ice wore away this divide and later, as the ice slowly receded back

257

to the Cortland region, the newly integrated valley became well-established as a meltwater channel. It was only natural, still later, for the Tioughnioga to seek this escape around the Valley Heads moraine blockage in the old Fall Creek valley. Thus the size of the postglacial, gouged-out valley is largely a function of the size and development of the preglacial valley. In this case, the presence of a bedrock divide between the headwaters of the two opposed streams also placed a limit on glacial erosion.

Numerous small glens may be seen where tiny tributary streams have notched the walls on either side of the Tioughnioga. Many new rock cuts opened up during road construction in recent years expose dark shales with interbedded siltstone and sandstone. Whitney Point is on a fairly broad flood and delta plain where the Otselic River joins the Tioughnioga. The Whitney Point Flood Control Dam impounds the Otselic just upstream from the village.

Between here and Chenango Bridge (13 miles), the road makes an overland traverse that gives a good impression of the topographic flavor of the uplands country above the major troughs. The relief here is somewhat subdued, and the hills have been smoothed and rounded by overriding ice, creating somewhat steeper north slopes and a lovely, rolling landscape.

From Chenango Bridge to Binghamton (2 miles), the highway follows the west side of the Chenango River, which joins the Susquehanna in the city. It is this confluence, in fact, that determined the location of the city.

The city of Binghamton is built upon a broad plain of glacial-river deposits that fill the Susquehanna Valley to a depth of 200 feet or more. The river flows westward from the city to Waverly (35 miles) where the Chemung River joins it from the west, and the main branch turns southward to Scranton, Pennsylvania, and beyond.

Between Binghamton and the Pennsylvania border, the highway follows the broad Susquehanna Valley, mostly high on the east wall, with good views over the river. The river now is much smaller than it was during Wisconsin deglaciation when, for a time, it carried nearly all of the copious meltwater from the Allegheny Plateau, including all of the juvenile Finger Lakes.

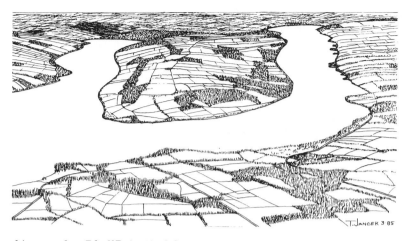

Looking north to Bluff Point and the two arms of Keuka Lake with Penn Yan at the north end of the right branch. The low area across the upper end of Bluff Point was submerged during deglaciation, making an island of the bluff.
Drawing by T. Jancek

NY 54:
Penn Yan—Bath
26 mi./42 km.

Keuka Lake is the only one of the Finger Lakes that branches. Canandaigua Lake would be similar if its water level were high enough to fill the branching West River trough (see NY 21, 371: Canandaigua—Cohocton discussion). In fact, the two lakes would practically be mirror images of each other and would look, from the air, like side views of a pair of hands with thumb and forefinger extended. The Y of each lake points south, indicating, with little question, that the troughs were hollowed out of south-flowing preglacial drainage systems. Other western lakes that occupy south-pointing branching valleys are Hemlock and Canadice. Keuka Lake is unusual in one other respect: it is the only one of the major Finger Lakes that drains directly into one of the other lakes. Keuka outlet passes through Penn Yan at the end of the east branch and flows eastward into the midsection of Seneca Lake at Dresden. The divide between the two lakes here is not very high, and Keuka Lake is nearly 300 feet higher than Seneca.

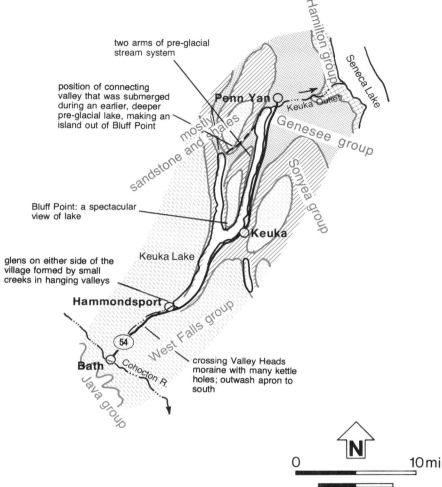

two arms of pre-glacial stream system

position of connecting valley that was submerged during an earlier, deeper pre-glacial lake, making an island out of Bluff Point

Bluff Point: a spectacular view of lake

glens on either side of the village formed by small creeks in hanging valleys

crossing Valley Heads moraine with many kettle holes; outwash apron to south

Penn Yan

Keuka Outlet

Seneca Lake

Hamilton group

Genesee group

Sonyea group

mostly sandstone and shales

Keuka

Keuka Lake

Hammondsport

West Falls group

54

Bath

Cohocton R.

Java group

0   10 mi

0   10 km

N

Keuka is the only Finger Lake in which two arms of a preglacial south-flowing stream system are now submerged.

**NY 54
PENN YAN—BATH**

Between Penn Yan and Keuka (11 miles), NY 54 stays close to the east shore of the east branch with numerous glimpses of the lake. On the way you pass over many deep gullies of small, lateral streams with large fan-shaped deltas on the lake. The gullies and a few roadcuts expose late Devonian Genesee group shales. Gullies revealed by tree stripes on the flank of Bluff Point across the lake are much smaller and shallower because they have less drainage area to draw from on the narrow divide between the two lake branches.

Keuka provides good views of Bluff Point. From the point itself, the panorama is one of the grandest of the Finger Lakes region. It encompasses parts of both branches of the lake and the "main trunk" to the south. One has the impression of standing on a lofty hill on an island. The description is apt because this prominence was actually an island during one proglacial lake stage when the water level was much higher than it is now. The two branches were interconnected across Bluff Point by a pass that now stands at an elevation of 823 feet, or just over 100 feet above the present water level. The highest point on the divide is 1426 feet, or about 700 feet over the lake.

The soft, rounded profile of Bluff Point is a product of glacial streamlining that is almost unchanged by modern erosion. The rock floor just south of the point is much deeper than in the troughs to the north, apparently because of the doubled "scour-power" of the combined ice from the two valleys.

The road continues close to the shore between Keuka and Hammondsport (8 miles). The convex roll of the trough walls becomes more apparent to the south, as they also get considerably higher. The sides rise nearly 1200 feet above the southern end of the lake.

Hammondsport is on the flat lake-and-delta plain at the southern tip of the lake. Two moderately large glens descend the slopes on either side of the village.

Between Hammondsport and Bath (7 miles), you pass over a segment of Valley Heads moraine with lots of small kettle holes. Bath is on an extremely wide, pitted outwash apron with numerous small kettle lakes at the junction of the Keuka trough with the Cohocton River valley.

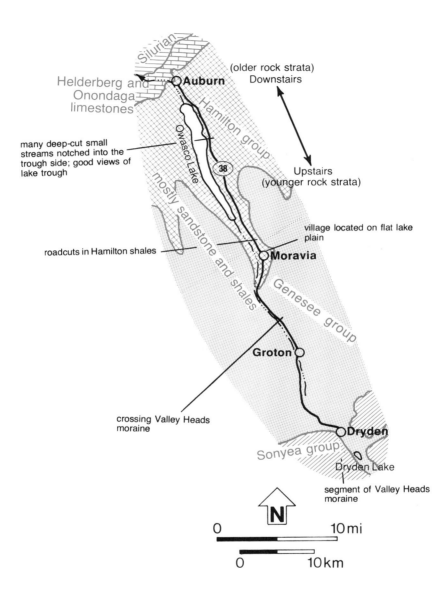

Silurian

Helderberg and
Onondaga
limestones

(older rock strata)
Downstairs

Auburn

Hamilton group

Owasco Lake

38

many deep-cut small
streams notched into the
trough side; good views of
lake trough

Upstairs
(younger rock strata)

village located on flat lake
plain

mostly sandstone and shales

roadcuts in Hamilton shales

Moravia

Genesee group

Groton

crossing Valley Heads
moraine

Dryden

Sonyea group

Dryden Lake

segment of Valley Heads
moraine

N

0   10 mi

0   10 km

NY 38
AUBURN—DRYDEN

NY 38:
Auburn—Dryden
39 mi./63 km.

This route follows the Owasco Lake trough from end to end and shows many of the same physical characteristics as NY 41 in the Skaneateles trough.

Auburn, at the north end of the lake, is on hummocky moraine deposits that have been breached by the north-flowing Owasco outlet. The moraine, of course, records a stand of Wisconsin ice later than Valley Heads. Between Auburn and Moravia (19 miles), the road follows the west wall of the trough, with good views over the lake, expecially from the high points near its south end. You cross over numerous, small glens en route, and more are revealed on the opposite wall of the trough by tree-lined stripes. Several roadcuts on the slopes north of Moravia expose grayish Hamilton group shales with strong vertical joint fractures.

Moravia lies on flat lake plain about 4 miles from the south end of the lake. The village is at the mouth of Decker Creek Glen and just north of Fillmore Glen State Park. Both creeks are tributary to north-flowing Owasco Inlet.

Between Moravia and Groton (10 miles), you follow the southern extension of the narrowing trough over an unusually long segment of Valley Heads moraine, which also has exceptionally well-developed delta terraces on the valley sides. The main street in Groton goes over a lump in the center of town that is the high point of the moraine. The moraine is now almost completely breached by Owasco Inlet, whose headwaters lie 5 miles south of Groton just short of Fall Creek near Freeville. It is interesting to compare this drainage pattern with that of the preglacial Fall Creek discussion in NY 13: Cortland—Horseheads. Then it was Fall Creek that theoretically extended itself to capture the previously established headwaters of the preglacial, north-flowing Owasco River. Now, the north-flowing Owasco Inlet is poised to capture Fall Creek. The breaching of the moraine is similar to that of Catherine Creek in the southern reach of the Seneca Lake trough discussed in NY 14: Geneva—Horseheads.

At Freeville, the road crosses Fall Creek and follows its tributary, Virgil Creek, between there and Dryden. Virgil Creek now occupies the former headwaters valley of the preglacial Owasco River. More moraine lies just south of Dryden and contains a large kettle lake called Dryden Lake.

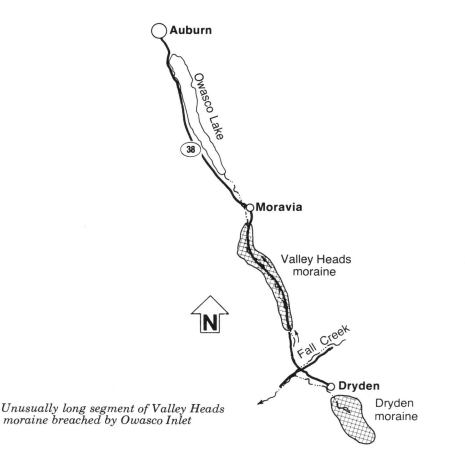

Unusually long segment of Valley Heads moraine breached by Owasco Inlet

*View north along
Skaneateles Lake.*
Courtesy
Kendall / Hunt
Publishing Co.

NY 41:
Skaneateles—Homer
27 mi./44 km.

NY 41 skirts the east shore of Skaneateles Lake and follows the southern end of the lake trough to its junction with the Tully trough at Homer. Following the trough as it does from one end to the other, the route offers excellent views—in a short distance—of the geologic characteristics that exemplify all of the Finger Lakes. Included is the lake itself, Valley Heads moraine and outwash plain, steep bedrock walls, truncated spurs, glens, and more.

The peculiar, long, low promontory on the west shore just south of Skaneateles is a partly drowned drumlin.

Between Skaneateles and Scott (20 miles), several roadcuts expose brownish middle Devonian Hamilton shales, and one shows Skaneateles limestone belonging in the same group. There are several lovely views of the lake, notably one by a parking area 2.8 miles north of Scott opposite the south end of the lake. On a clear day it is possible to survey nearly the entire 15-mile length of the lake. This is certainly one of the most beautiful of the Finger Lakes, especially along its southern reach where the green-carpeted slopes are steepest and the lake exceptionally narrow. Skaneateles water is unusually clear and has a distinctive, pale greenish hue that differs markedly from that of most of the other lakes.

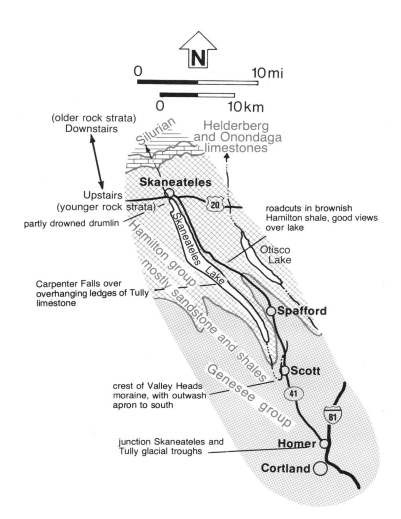

(older rock strata)
Downstairs

Helderberg
and Onondaga
limestones

Silurian

Skaneateles

Upstairs
(younger rock strata)

partly drowned drumlin

Hamilton group

mostly sandstone and shales

Skaneateles Lake

roadcuts in brownish
Hamilton shale, good views
over lake

Otisco
Lake

Carpenter Falls over
overhanging ledges of Tully
limestone

Spafford

Genesee group

Scott

crest of Valley Heads
moraine, with outwash
apron to south

junction Skaneateles and
Tully glacial troughs

Homer

Cortland

NY 41
SKANEATELES—HOMER

266

Numerous small creeks crease both trough walls. Their deltas constitute the only irregularities of the otherwise smooth shorelines. The largest glen is on the west side near the hamlet of New Hope, where Carpenter Falls cascade over an overhanging ledge of resistant Tully limestone at the head of the canyon. The Tully, at the base of the Genesee group and supported by Hamilton shales, is a thin, but persistent, ledge-former throughout the eastern Finger Lakes region.

The Valley Heads moraine crests at about 435 feet above lake level one mile west of the hamlet of Scott, near the point where the moraine ends and the outwash apron begins. The through-valley divide is now almost breached by Grout Brook that flows south through Scott and then turns sharply north to the lake, cutting deeply into the moraine. The position of the moraine is in line with other segments of the Valley Heads in the Otisco and Tully valleys to the east.

The narrow southern extension of the trough between Scott and Homer (7 miles) is deeply buried in outwash, kame, and perhaps also glacial lake sediments, resulting in a flat floor and abrupt, moderately steep sides. Over much of the way there is a conspicuous kame terrace on the west side. The west wall is also creased by two deep glens, called Vrooman Gulf (northern) and Homer Gulf.

Homer is on the broad, flat sediment plain at the Y-junction between the Skaneateles trough and the Tully trough. The Tully was one of the main sluiceways for glacial meltwaters.

The narrowness of the southern reach of the Skaneateles trough suggests that this was near the headwaters valley for the preglacial, north-flowing, Skaneateles River, which was widened and deepened by glaciation and whose flow direction was reversed. Theoretically, the headwaters, which were still farther south, were pirated by Fall Creek.

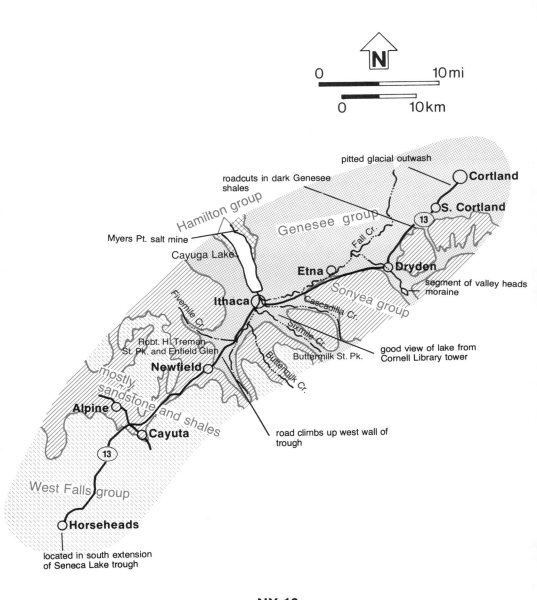

0   10 mi

N

0   10 km

pitted glacial outwash

roadcuts in dark Genesee shales

Cortland

Hamilton group

Genesee group

S. Cortland

13

Myers Pt. salt mine

Fall Cr.

Cayuga Lake

Etna

Dryden

segment of valley heads moraine

Ithaca

Sonyea group

Cascadilla Cr.

Fivemile Cr.

Sixmile Cr.

Robt. H. Treman St. Pk. and Enfield Glen

good view of lake from Cornell Library tower

Newfield

Buttermilk Cr.

Buttermilk St. Pk.

mostly sandstone and shales

Alpine

Cayuta

road climbs up west wall of trough

13

West Falls group

Horseheads

located in south extension of Seneca Lake trough

NY 13
CORTLAND—HORSEHEADS

NY 13:
Cortland—Horseheads
52 mi./84 km.

This route follows the bedrock valley of the preglacial Fall Creek southwestward between Cortland and Ithaca (23 miles). Two major stream systems instead of one presently occupy the valley: upstream from Cortland are the east and west branches of the Tioughnioga River, which join by the city and turn east; downstream, 7 miles from Cortland, the modern Fall Creek enters from the north and continues downvalley to the Cayuga Lake trough at Ithaca. Separating the two is a vast deposit of Valley Heads moraine.

The Fall Creek valley fill extends for at least 13 miles downvalley from Cortland. At the city, the surface is quite flat, and the valley sides to south and west are marked by well developed kame terraces. Apparently, most of the fill consists of fairly clean, clay-poor, gravelly kame and outwash deposits, with a subordinate amount of till. Downvalley from the city the surface becomes pitted, dotted with a few small kettle lakes and marshy hollows, and is more mounded. There is a cluster of kame mounds at Malloryville (7 miles from Cortland) with a small esker ridge. The modern Fall Creek has apparently destroyed much of the original esker, which may have extended at least three miles farther downvalley.

The highway actually goes overland between South Cortland and Dryden, climbing up the southeast wall of Fall Creek Valley, with open views of the rolling upland country. En route you pass a few cuts in dark shales of the late Devonian Genesee group. Dryden is in

Virgil Creek Valley, a tributary to Fall Creek. Beginning about a mile southeast of the village and stretching for at least 5 miles, the valley is plugged with yet another segment of Valley Heads moraine with Dryden Lake perched in it.

The road goes westward between Dryden and Etna (5 miles), and returns to the south wall of Fall Creek Valley; this is the Portage scarp mentioned in *Language of the Landscape*. Between Etna and Ithaca (6 miles), you descend about 700 feet, with lovely panoramic views down Fall Creek Valley and into the Cayuga trough. This lower section of Fall Creek is deeply incised into the valley fill. In its last mile or so before issuing onto the Cayuga floor, it has cut a picturesque deep gorge, with numerous waterfalls and rapids, into the Genesee bedrock of the trough wall. The gorge marks the north boundary of the Cornell University campus. The south edge of the campus is at another, but smaller, gorge of Cascadilla Creek. Incidentally, one of the best places to get an overview of Cayuga Lake, the troughs to the south, and the surrounding landscape of the Allegheny Plateau is from the Baker Library Tower at Cornell.

The city of Ithaca is mostly built on a huge, composite, trough-filling delta constructed by Fall, Cascadilla, Sixmile, and Cayuga inlet creeks. The delta is more than 400 feet thick and continues to build out into the lake today, gradually displacing its south shore northward.

The Ithaca area probably has a greater number and variety of deep bedrock glens than any other section of the Finger Lakes region, and certainly most of the ones developed for tourism. A few miles north of the city is Taughannock Falls State Park, with its 215-foot-high falls, the highest in the state. Just south of the city on NY 13, you pass Buttermilk Creek Glen of Buttermilk Falls State Park on the east side of the trough, and Coy Glen on the opposite side. About 2 miles farther south is Enfield Glen in Robert H. Treman State Park where the highway begins to climb out of the west side of the trough. These are only a few of the more spectacular gorges but numerous others are undeveloped.

Between Enfield Glen and Newfield (3 miles), the road climbs more than 600 feet up the west wall of the trough, with excellent open views south over the Valley Heads moraine, with its hummocky surface of knobs and kettles littered with boulders. This is undoubtedly the largest and most representative segment of the moraine, recognizable for about 12 miles along the valley floor, from Coy Glen to North Spencer. The northern 2-3 miles is eroded and largely buried under glacial lake and deltaic sediments, whereas the southern end, in the usual fashion, grades rapidly to pitted outwash. At its highest,

270

Enfield Glen, showing strong joint (fracture) control of stream courses, cut in upper Ithaca shales and lower Enfield shales and sandstones (Devonian)
Courtesy Kendall / Hunt Publishing Co.

the moraine is about 750 feet above the present level of Cayuga Lake, so there is no doubt as to its effectiveness as a dam, not only for the present lakes, but also for some of its much higher, proglacial predecessors. At the earliest and highest levels, when the ice front still blocked the trough to the north, meltwaters overtopped the moraine and flowed southward to the Susquehanna River.

The hamlet of Newfield is in branching Pony Hollow trough astride more Valley Heads moraine. A large excavation nearby reveals the moraine's till as a rather chaotic mixture of sand and gravel with little of the very fine clay often found in moraine deposits.

Southwest of Newfield, the rest of the way to Alpine Junction (8 miles), you cross 2 miles of moraine and then follow the remarkably flat outwash apron of the Pony Hollow trough with its dramatically steep walls, especially on the north side. Nowhere is this distinctive trough profile more apparent than in the area around Cayuta where Pony Hollow widens and joins with the Cayuta Lake outlet trough.

271

The modern Cayuta Lake is a small residual of glacial Lake Cayuta that occupies a basin 6 miles north-northwest of Cayuta and the same distance east of Watkins Glen. The lake now drains southward by way of Cayuta Creek through a bedrock gorge called The Gulf. It then follows the outlet trough to Cayuta and beyond, eventually joining the Chemung at Waverly on the Pennsylvania border. An interesting 700 foot deep bedrock gorge, called the Hendershot Gulf, is about a mile north of Cayuta. This canyon is aligned with The Gulf and an intermediate gorge. All three were carved along the ice margin by overflow from proglacial Lake Cayuta during a time when the ice still blocked the Cayuta Creek valley slightly to the west of them.

Between Alpine Junction and Horseheads (12 miles), you go overland to the southern extension of the Seneca Lake trough. The valley and hills around Horseheads are a virtual museum of glacial features that characterize the Finger Lakes region. These are discussed in NY 17: Binghamton—Arkport/Hornell. Suffice it to say here that the trough at Horseheads is an extremely broad, flat, pitted, outwash plain, with the Valley Heads moraine beginning just north of the city, and that major drainage changes occurred here as a result of Wisconsin deglaciation.

X

RIDGE ROAD
AND THE SEAWAY TRAIL

Lewiston is at the mouth of the Niagara Gorge, right at the base of the Niagara scarp. This peaceful and unassuming place conceals secrets of one of the most dramatic events in the geologic evolution of the modern New York landscape. It was here that Niagara Falls were born approximately 12,000 years ago in the waning stages of the Ice Age.

The Wisconsin ice sheet, in its slow northward recession, had just backed off of the scarp and waters of the newly-forming upper Great Lakes began spilling over the scarp into juvenile Lake Iroquois. The glacier had left behind one extra lake, called Lake Tonawanda, that filled a shallow depression in the Tonawanda plain between the Niagara and Onondaga scarps and between Niagara Falls and Rochester (see US 20: Buffalo—Canandaigua discussion). Lake Tonawanda, thus, served as an intermediate basin one step down from Lake Erie and one step above Lake Ontario. Tonawanda initially poured over the scarp not just at Lewiston, but also at four other widely separated localities to the east: Lockport, Gasport, Medina, and Holley (see map on page 42). Because of a slight westward tilt of the surface, however, Lewiston received the lion's share of the discharge, and eventually Niagara Falls were able to pirate all of the flow away from the other spillways before they had carved much of a slot in the scarp face. That happened because of the gentle southward slope of the late Silurian Lockport dolomite caprock of the Niagara scarp.

With its greater initial flow, Niagara Falls cut more rapidly down the dip of the caprock than the other falls, lowering the spillway level faster, which in turn controlled the height of the lake. When the lake fell below the other spillway levels, those falls simply dried up. When the level fell even further, the lake emptied. Niagara Falls took over and has been in control ever since.

The Niagara scarp has played an extremely important role in the geomorphic history of New York. It is more than just a local feature, for it extends almost continuously around Lakes Huron and Michigan into northern Illinois. In western New York, it trends roughly parallel to the Ontario shoreline and diminishes in height to the east. The scarp is almost imperceptible east of Rochester, where it appears to have been most severely scoured by ice pushing out of the Ontario basin. An excellent place to view the scarp in profile is from the slopes of the Earl W. Brydges Artpark behind Lewiston, especially when looking west to the Canadian side of the river.

NY 104:
Lewiston—Rochester
70 mi./113 km.

THE RIDGE ROAD

Lewiston lies in the shallow plunge basin of the original Niagara Falls, which, thus far, have migrated 7 miles upstream to the south. From Lewiston, the river flows northward for another 7 miles to enter Lake Ontario, crossing the Lake Iroquois plain over Queenston formation red shales of late Ordovician age.

Between Lewiston and Dickersonville (5 miles), NY 104 traverses the Irondequoit limestone shelf a step down from the main Lockport scarp that hovers nearby. The Irondequoit and the underlying Reynales limestone together form a thin resistant ledge under the falls, throughout the gorge, and in the face of the scarp that is separated from the overlying Lockport by a relatively thick section of recessive Rochester shale (see Niagara Falls discussion).

The remainder of the route all the way to Rochester follows the lake plain, with the scarp almost entirely out of sight to the south. It is a route of unique geological interest because it lies atop the beach ridge that delineates the shoreline of former Lake Iroquois; hence the name Ridge Road. The ridge is barely perceptible west of Lockport, but it builds up to a more conspicuous crest farther east. It is nowhere a very prominent topographic feature. Over most of its length, the ridge just banks off gently and evenly to the north, blending into the lake plain, whereas the south side is bankless and relatively uneven, with a

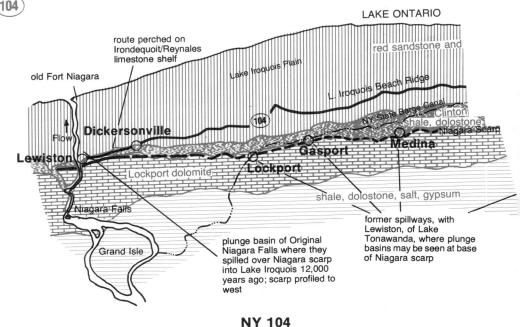

LAKE ONTARIO

route perched on
Irondequoit/Reynales
limestone shelf

red sandstone and

old Fort Niagara

Lake Iroquois Plain

L. Iroquois Beach Ridge

NY State Barge Canal

Clinton

Dickersonville

Flow

104

shale, dolostone

Niagara Scarp

Lewiston

Gasport

Medina

Lockport dolomite

Lockport

Niagara Falls

shale, dolostone, salt, gypsum

Grand Isle

plunge basin of Original
Niagara Falls where they
spilled over Niagara scarp
into Lake Iroquois 12,000
years ago; scarp profiled to
west

former spillways, with
Lewiston, of Lake
Tonawanda, where plunge
basins may be seen at base
of Niagara scarp

NY 104
LEWISTON—ROCHESTER

gently rising slope. In some places it is like a railroad grade with banks on both sides and mucky land to the south. This suggests an origin as an offshore bar, with a lagoon behind it, like those of the south shore of Long Island. In the eastern part, the hummocky, pitted landscape of the Rochester-Albion moraine belt lies just to the south.

The ridge is made up of sandy, gravelly deposits derived largely from glacial debris reworked by wave action along the Iroquois shore in exactly the same way that Lake Ontario waves do today. Later, when the St. Lawrence valley was free of ice, allowing Lake Iroquois to drain through it, the lake shrank rather quickly, leaving the ridge high and dry as you see it today.

No account of Ridge Road would be complete without mention of its many lovely, old cobblestone homes, churches, schoolhouses, and other buildings which reveal a human history equally as unique as, and a product of, the geological history. Practically all of the structures were built in the period from 1825-1860, after completion of the Erie Canal, which lies a few miles to the south. Completion of the canal left a lot of skilled masons unemployed in this region at about

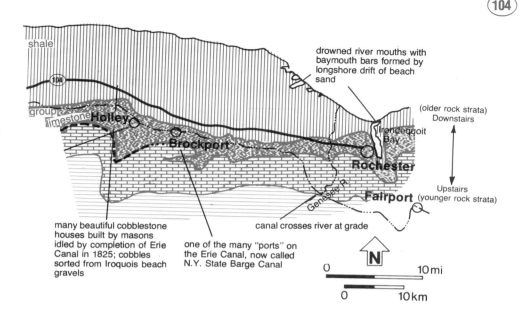

drowned river mouths with baymouth bars formed by longshore drift of beach sand

shale

group limestone

Holley

Brockport

Irondequoit Bay

Rochester

Genesee R.

Fairport

(older rock strata) Downstairs

Upstairs (younger rock strata)

many beautiful cobblestone houses built by masons idled by completion of Erie Canal in 1825; cobbles sorted from Iroquois beach gravels

one of the many "ports" on the Erie Canal, now called N.Y. State Barge Canal

canal crosses river at grade

N

0   10 mi

0   10 km

the same time the early settlers were clearing the land. In search of work, the masons offered to build permanent homes for the farmers out of the abundant gravels from the surrounding fields. It became a highly competitive business, with the skills of the many masons pitted against each other. The net result is a variety of patterns and an unusually high level of craftmanship.

Cobblestone house on Ridge Road (NY 104)

There are a number of beautiful cobblestone patterns in the buildings along the Ridge Road, but nearly all of them have one distinguishing characteristic: the stones are arranged in neat horizontal rows. In the earliest buildings, stones of various sizes, shapes, and color were used; and the horizontal mortar joint was recessed so that stones projected. Later, the masons became much more selective, choosing stones of uniform size, shape, and color. Many even employed a template like those used for grading eggs, to ensure the uniformity of stone size. Masons vied with each other in the use of patterns and the embellishment of mortar joints.

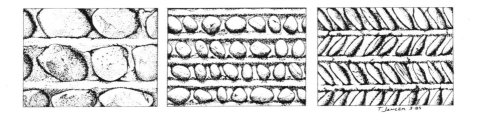

Patterns typical of the cobblestone houses of Ridge Road, including: 1) larger, variable cobbles in the earlier houses; 2) smaller cobbles of more uniform size and shape in later houses; and 3) herringbone pattern. A common feature was the arrangement in horizontal rows. The beaded mortar separating rows was a later decorative feature. Drawing by T. Jancek

One of the most intriguing patterns is the herringbone, in which elongate ovoid cobbles were laid up in one row slanting to the right, and in the next slanting to the left, and so on. Some walls were laid up with stones all of the same color. In the later buildings, the recessed mortar gave way to "beaded" mortar, which makes individual rows, or even individual stones, look as if they were framed. The importance of the mortar needs to be emphasized. What the masons did with it certainly enhanced their designs; even more significant, however, is the high quality of the mortar used, for that is the real secret of preservation of these many beautiful walls today.

NY 104:
Rochester—Oswego
74 mi./120 km.

In Rochester, NY 104 crosses the Genesee River within view of the lower falls of the Rochester Gorge. The city of Rochester was established by the river because of the availability of cheap water power for use in operating flour mills. At the time, the fertile southern tier of New York was one of the world's largest wheat producing regions. So many flour mills were once operating in the city that it was dubbed the "Flour City." When the flour industry died out, many nurseries and parks were developed; and the city's nickname was changed to "Flower City."

The rocks exposed in the streambed beneath the Genesee bridge are redbeds of the late Orodovician Queenston formation, which is known to underlie much of Lake Ontario and have a full subsurface thickness of about 1000 feet (see Niagara Frontier chapter). The lower falls cascade 100 feet over the Thorold sandstone of the early Silurian Clinton group, which is separated from the Queenston by Grimsby formation redbeds. Farther upstream, at the head of the gorge, the upper falls plunge over the massive, resistant Lockport dolostone beds. The stratigraphic section throughout is almost identical to that of the Niagara Gorge.

Between the Genesee bridge and Irondequoit Bay, the road continues along the Lake Iroquois beach ridge, with lake plain falling away gradually to the north and more rugged topography to the south. Four miles to the south are the Pinnacle Hills, a moraine-kame dump pile that marks a stand of the receding Wisconsin ice front.

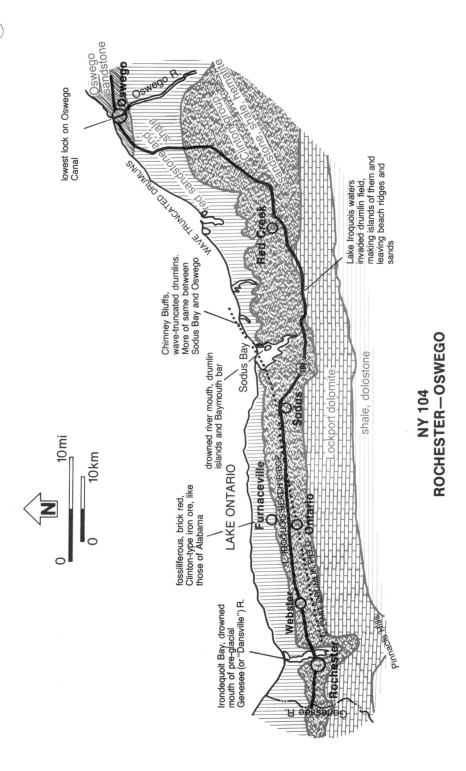

NY 104
ROCHESTER—OSWEGO

280

When these hills were formed, the region south of them was inundated by a proglacial meltwater lake called Lake Dana that drained eastward to the Mohawk Valley. Further retreat of the ice permitted a new drainage route to the north across a low point in the Pinnacle Hills, thus initiating the modern course of the Genesee River.

Irondequoit Bay is the drowned mouth of the preglacial Genesee River (see Letchworth chapter). The old course veered east of the modern north-flowing one from 15 miles south of Rochester and continued for 13 miles to the present site of the hamlet of Fishers. There it turned north again to the "Ontarian River," the preglacial, east-flowing river that occupied the Ontario basin. The lower valley now contains tiny Irondequoit Creek that flows into the bay. The mouths of nearly all of the streams along the south shore of Lake Ontario are drowned because glacial rebound tilted the Ontario basin slightly to the south. Baymouth bars, like the one visible north of the

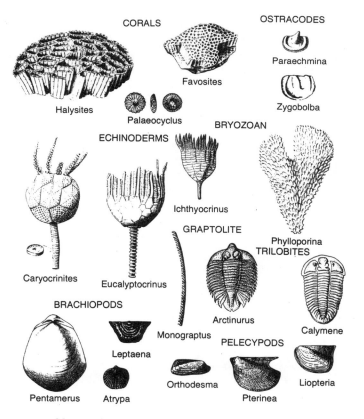

Representative Clinton fossils drawn at scales of .3 to .7 times normal size, except for the ostracodes in upper right, which are 5 to 10 times normal size.
Courtesy Kendall/ Hunt Publishing Co.

bridge across Irondequoit Bay, are ubiquitous, a result of the long-shore drift of beach sands.

The Iroquois beach ridge continues east of Irondequoit Bay to Sodus, but NY 104 lies just north of it most of the way. The ridge is best defined around Webster, 5 miles east of the bay. North of the village of Ontario, 8 miles east of Webster, are the old digs of the Furnaceville iron ores. The brick red ore occurs in the Furnaceville hematite member of the Reynales limestone, belonging to the early Silurian Clinton group. Similar "Clinton-type" iron ores are the basis for a major steel industry in Birmingham, Alabama. They are essentially limestones that have been replaced by the iron oxide mineral hematite in the peculiar, widespread, highly oxidizing environment of middle Silurian time. Many of the ores, including those of Furnaceville, are fossiliferous, with individual fossils completely replaced by hematite while they retain their original forms.

The line of the beach ridge becomes more irregular near Sodus, 12 miles east of Ontario, as it merges with the northern edge of the great drumlin field of western New York. This stretch of highway offers an excellent opportunity to see the transition from lake plain to drumlinoid topography. The drumlins at Sodus are truncated at Ridge road by Lake Iroquois wave action in exactly the same way that many drumlins along the present Ontario shore between Sodus Bay and Oswego are being cut back today.

7.5-minute topographic map of Sodus Bay, showing drumlin islands and wave eroded drumlin at Chimney Bluff (upper right)
Courtesy Kendall / Hunt Publishing Co.

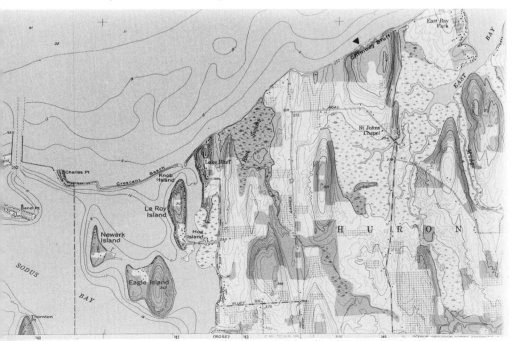

Badlands erosion of drumlin at Chimney Bluff State Park
Courtesy Kendall / Hunt Publishing Co.

Near the highway between Sodus and Red Creek (21 miles), Iroquois waters spread widely among the drumlins, making them into islands. They are now recognized as such by their wave-eroded sides and the surrounding marshy lake plain. Sodus Bay and nearby Chimney Bluff State Park are excellent places to see the modern counterparts of these features.

Between Red Creek and Oswego (21 miles), the Iroquois shoreline is obscure, as the road rolls over the heart of the drumlin field. At Oswego, you cross the Oswego River and Oswego Barge Canal. The canal is still extensively used for commercial purposes, especially by the cement block companies. The barges can carry 1500-1600 tons, several times the capacity of a semi-tractor-trailer, and the charge per ton is much lower. The canal connects with the New York Barge, or Erie, Canal at Three Rivers, 22 miles straight southeast of Oswego. Seven locks span the vertical distance of 118 feet, with the highest lift of 25 feet at Fulton.

Lowest lock on the Oswego Canal at Oswego, alongside the Oswego River

Detail of lower Genesee River and its preglacial route through Irondequoit Bay, the drowned river mouth

See map page 370.

NY 104, NY 104B, NY 3:
Oswego—Watertown
58 mi./ 94 km.

THE SEAWAY TRAIL

This route curves around the southeast corner of Lake Ontario and then follows the shore north to Sackets Harbor, where it turns east for a few miles to Watertown. It skirts the low, dip slope side of the Tug Hill tilted mesa on a landscape of very low relief. There are virtually no outcrops, but the route crosses several formation boundaries that are revealed to some extent by subtle scarps and divisions between both well and poorly drained areas.

The most conspicuous topographic features are the drumlins, but none are as well-formed nor as high as those of the great drumlin field farther west. The best developed drumlins on this route are along NY 104 and NY 104B east of Oswego (18 miles), where they have a consistent south-southeast alignment. In the 12 miles north of the NY 3 junction, the drumlins have more rounded outlines and are scattered. Yet farther north, they are again elongate and indicative of a northeasterly ice movement out of the Ontario basin. The radial trend from Lake Ontario thus indicated continues westward to Buffalo; it shows that the lake basin served as a spreading center for Wisconsin ice.

Certainly the greatest geologic and scenic attraction of the Seaway Trail is the shore itself, which, though visible from only a few points on the road, is easily reached from it. The shore is dotted with historic sites, public campsites, beaches, and recreation areas, including from south to north Old Fort Ontario in Oswego, the Progress Center, Pleasant Point Beach, Selkirk Shores, Brennan Beach, Rail City, Southwick Beach, Westcott Beach, and Sackets Harbor Battlefield. Between Oswego and Henderson Bay, the shore is similar in many details to the south shore of Long Island, with drowned river mouths forming lagoons behind a smooth, curving line of barrier bars.

You cross the Salmon River west of Pulaski. Farther upstream, where it passes through the village, the river has carved an impressive small gorge into the middle Ordovician Pulaski formation consisting of interbedded shales and siltstones. The beds are stacked like a layercake and sliced through by intersecting vertical joints. This arrangement, coupled with the entrenched meandering course and steep gradient, makes for some interesting rapids that are a favorite among whitewater enthusiasts. Each spring there is a whitewater derby here, when the two bridges in town and the fishery farther upstream are jammed with spectators and the river is full of colorful kayaks and canoes.

Between Henderson and Watertown (20 miles), subtle, low, tilted mesas are carved from the Trenton limestone beds as erosional outliers of the Tug Hill. These are highest and most conspicuous around Watertown, where they constitute a giant flight of stairs at the north end of the plateau (see Interstate 81: Syracuse—Alexandria Bay discussion).

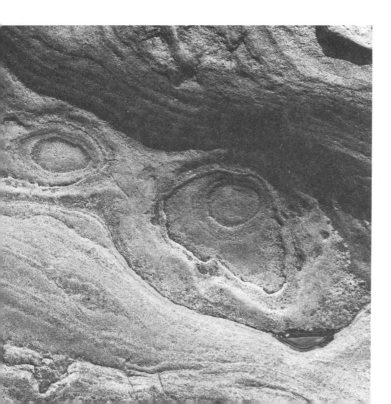

Stream erosion in gneiss

NY 37:
Malone—Ogdensburg
78 mi. / 126 km.

Between Malone and Ft. Covington (14 miles), NY 37 descends gradually from 700-plus feet to 180 feet above sea level, as it follows the gentle dipslope of the lower Paleozoic formations that floor the St. Lawrence Lowlands (see the Old Military Turnpike discussion). The general impression looking northward, especially from points near Malone, is that you are approaching the flat surface of the sea. According to present theory, the strata dip in this direction is a result of the doming of the Adirondacks, an event which began a few million years ago, and is still going on today at a rate of about 3mm/year. The beds appear to have once blanketed the Adirondack region. Erosion initiated by the uplift has stripped them from the mountains and tilted them upwards around the edges. On the human clock, 3 mm/year seems insignificant, but on the earth clock, it is a very rapid rise that far exceeds the rate of erosion. Over the geologically short span of a million years this would lead to an elevation of 3 kilometers!

The flat landscape is punctuated by till hills, low smoothly rounded mounds of glacial sediments. Numerous sand and gravel pits near the road also reveal the presence of shore and near-shore deposits formed in Lake Iroquois and the Champlain Sea during Wisconsin deglaciation. At Malone, the bedrock is Potsdam sandstone, exposures of which are visible in the small gorge of the Salmon River in town. Contact with the overlying Theresa formation sandy dolostone lies at about 5 miles north of Malone. The uppermost Theresa is exposed in

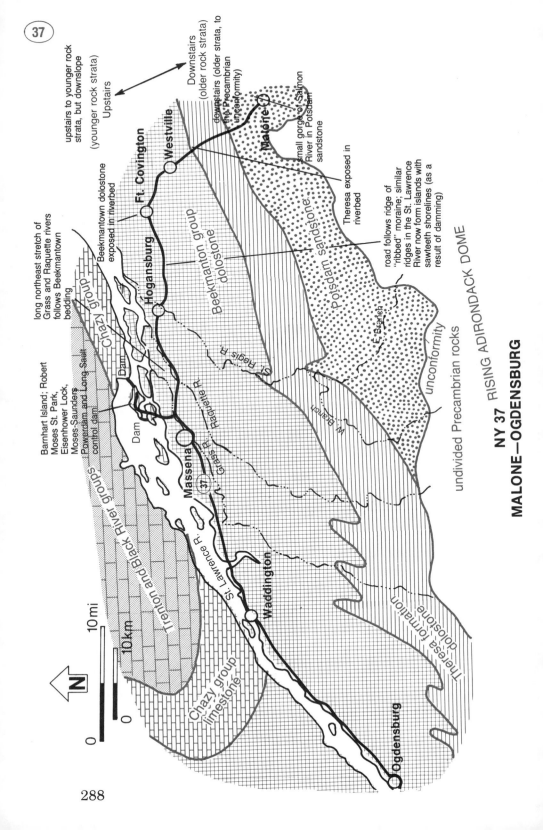

upstairs to younger rock strata, but downslope

Upstairs
(younger rock strata)

Downstairs
(older rock strata)

downstairs (older strata, to the Precambrian unconformity)

Malone

small gorge of Salmon River in Potsdam sandstone

Theresa exposed in riverbed

Westville

road follows ridge of "ribbed" moraine; similar ridges in the St. Lawrence River now form islands with sawteeth shorelines (as a result of damming)

Ft. Covington

long northeast stretch of Grass and Raquette rivers follows Beekmantown bedding

Beekmantown dolostone exposed in riverbed

Beekmantown group dolostone

Potsdam sandstone

Hogansburg

Chazy group

E. Branch

unconformity

undivided Precambrian rocks

RISING ADIRONDACK DOME

Dam

St. Regis R.

Barnhart Island; Robert Moses St. Park, Eisenhower Lock, Moses-Saunders Powerdam and Long-Sault control dam

Trenton and Black River groups

Dam

Raquette R.

Massena

Grass R.

37

St. Lawrence R.

Waddington

Chazy group limestone

Theresa formation dolostone

Ogdensburg

10 mi

10 km

N

0

0

NY 37 RISING ADIRONDACK DOME

MALONE—OGDENSBURG

the Salmon River bed at Westville Center and contact with the overlying Beekmantown dolostones is nearby. Thus, as you go down the dipslope, you are also going up in the stratigraphic section, from older to younger beds. Still farther north, at Ft. Covington, the stratigraphically higher Beekmantown dolostone may be seen in the Salmon streambed.

Between Ft. Covington and Hogansburg (10 miles), the road follows a line of glacial moraine ridges, with others visible to the north and south. Several of these are "ribbed," i.e. the ridges trend more or less east-west, but their tops have cross ridges, or flutings, that trend about north-south. The ridges have been interpreted as till deposits that were molded by overriding ice, and that may be transitional to drumlins. The moraines are elongated transverse to ice advance, and the ribbing and drumlins parallel to it. The road climbs over two or three of these low ridges between Ft. Covington and Hogansburg.

At Hogansburg, you cross over the St. Regis River, with excellent exposures of Beekmantown, or Ogdensburg, dolostone in the streambed. Here it is thick-bedded, contains lenses of hard chert that weather in relief, and is cut by abundant vertical joints that have been widened by solution. It also displays numerous small solutional pockets, called vugs, that are lined with crystals of calcite, dolomite, and other minerals, a typical feature in this formation.

Most of the way between Hogansburg and Massena (11 miles), the road follows the Raquette River, which here is entrenched approximately 50 feet. It crosses over the river at Raquette; between there and Massena it follows the south bank of the Grass River. These two rivers are 10 miles apart where they pass through Canton (Grass) and Potsdam (Raquette), but here they join the St. Lawrence River at nearly the same point after running side-by-side for 13 miles.

Near Massena you pass the entrance to Robert Moses State Park on Barnhart Island and the Eisenhower Locks of the St. Lawrence Seaway. Several sites you may wish to visit on the island include the Long Sault Spillway Dam and the Moses-Saunders Power Dam. Flooding of the river has introduced profound changes in the shoreline and islands between here and Ogdensburg, and has submerged the International Rapids. Many of the islands north of Massena have sawteethed shorelines that result from flooding of ribbed moraine like those discussed above. The topography all around Massena, in fact, is almost entirely made up of ribbed moraine and drumlins. The hilliness is most apparent on the city bypass, where maximum relief is about 100 feet. At the end of the bypass, at the junction with NY 56, you are at the crest of one of these till hills at 281 feet. The road to the west drops down to cross the Grass River.

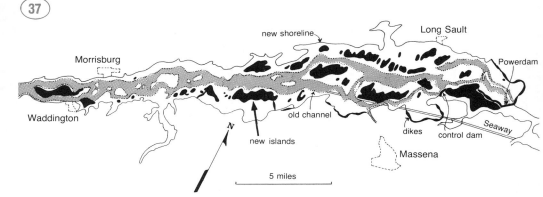

*Massena-Waddington section of Lake St. Lawrence behind the
Moses-Saunders and Long Sault dams, also showing the original shoreline.
Drowning of ribbed-moraine produced the sawteethed shorelines of some of
the new islands.*

Between Massena and Ogdensburg (34 miles), the route continues
close to the St. Lawrence River over numerous moraine and ribbed
moraine ridges and drumlins, but the relief is very low. Views en
route of the St. Lawrence River are frequently open, and it is quite
common to see ocean-going ships on their way up or down the river.
On the north side of Waddington, you cross Coles Creek, one of
numerous inflowing streams drowned by flooding of this segment of
the St. Lawrence River behind the dams. As noted above, nearly all of
these streams are entrenched, so that when the water level was
raised, embayments reached far upstream into the side valleys.
These effects diminish toward the southwest and are negligible at
Ogdensburg. Virtually no bedrock is exposed on or near the highway
from Massena to Ogdensburg. A lot of wet bottomland is in the
poorly-drained, locally clay-rich soil of glacial and glacial lake origin.

NY 37, NY 12, NY 12E:
Ogdensburg—Cape Vincent—
Watertown
87 mi./141 km.

At the south side of Ogdensburg, there is a large quarry in the Ogdensburg, or Beekmantown dolostone. Between Ogdensburg and Alexandria Bay (34 miles), there are numerous roadcuts that expose, from northeast to southwest: the Ogdensburg dolostone, then Theresa formation, Potsdam sandstone, and finally Precambrian gneisses. Thus the stratigraphic section progresses from younger to older rocks toward Alexandria Bay. The many large, new cuts along the route can only be described as spectacular. The geologic config-uration is similar to the approach to the Adirondack Dome from the St. Lawrance River described in the Old Military Turnpike section, but here you are "climbing up" onto another uplifted feature called the Frontenac Arch. The Arch trends northwestward through Alexandria Bay. Its uplift appears to be contemporaneous with dom-ing of the Adirondacks, and the apparently once continuous blanket of Potsdam and Theresa strata has been similarly removed by erosion on the crest of the uplift. The dip of these beds away from the Arch is almost imperceptible in single roadcuts, but it is indisputable. The stratigraphic thickness of each unit apparently doesn't change much along this route, and the level of the road itself is fairly constant. Therefore, the road has to trend up the dipslope of the beds to the southwest. Several cuts in the Theresa display broad, open folds and small fault displacements that may have been caused by the uplift.

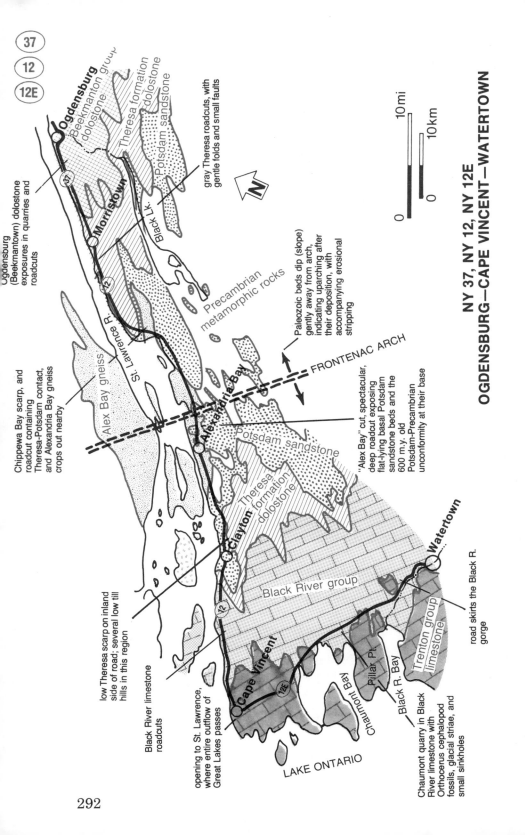

Ogdensburg

(Beekmanton) dolostone group

Theresa formation dolostone

Potsdam sandstone

gray Theresa roadcuts, with gentle folds and small faults

Ogdensburg (Beekmantown) dolostone exposures in quarries and roadcuts

Morristown

37

12

Black Lk.

Precambrian metamorphic rocks

Paleozoic beds dip (slope) gently away from arch, indicating uparching after their deposition, with accompanying erosional stripping

FRONTENAC ARCH

St. Lawrence R.

Alex Bay gneiss

Chippewa Bay scarp, and roadcut containing Theresa-Potsdam contact, and Alexandria Bay gneiss crops out nearby

Alexandria Bay

Potsdam sandstone

Theresa formation dolostone

"Alex Bay" cut, spectacular, deep roadcut exposing flat-lying basal Potsdam sandstone beds and the 600 m.y. old Potsdam-Precambrian unconformity at their base

Clayton

12

low Theresa scarp on inland side of road; several low till hills in this region

Black River limestone roadcuts

Black River group

Watertown

road skirts the Black R. gorge

Trenton group limestone

opening to St. Lawrence, where entire outflow of Great Lakes passes

Cape Vincent

12E

Pillar Pt.

Black R. Bay

Chaumont Bay

Chaumont quarry in Black River limestone with Orthocerus cephalopod fossils, glacial striae, and small sinkholes

LAKE ONTARIO

N

10 mi

10 km

0

0

NY 37, NY 12, NY 12E
OGDENSBURG—CAPE VINCENT—WATERTOWN

The road hugs the river shore most of the way, with numerous open views that often include ocean-going ships. Near Ogdensburg you pass over gently undulating land built upon till hills like those discussed in the preceding Malone-Ogdensburg section of NY 37. The Ogdensburg dolostone is present only on this side of the Arch, and the younger Black River and Trenton limestones only appear on the southwestern side. This suggests that the region of the Arch must have posed a depositional barrier during Ordovician time, first limiting Ogdensburg dolostone to the northeastern side, then limiting Trenton deposits to the southwestern side. The Ogdensburg dolostone-Theresa formation contact is somewhere between Ogdensburg and Morristown.

Chippewa Bay roadcut showing contact between Potsdam sandstone (white beds) and overlying sandy dolostone (gray beds) of the Theresa formation

One of the most significant roadcuts is at Chippewa Bay between the Cedar Point overlook and the Hamlet of Chippewa Bay. This long cut goes downhill through a scarp, exposing the lower beds of the Theresa and the upper beds of the Potsdam. The Theresa in this section consists of thin-bedded, gray, limey or dolomitic sandstone that contains fossil worm burrows, grazing trails, and broken shell material. The 20-foot thickness of Potsdam beds exposed at the lower

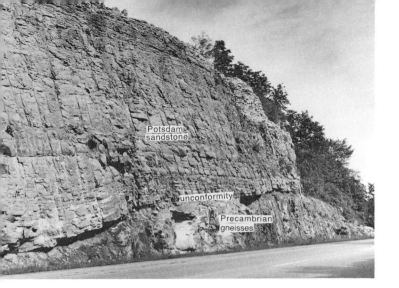

Alex Bay cut, an angular unconformity between Precambrian gneisses (1.1 billion years old) and Cambrian Potsdam sandstone (500 million years old)
Courtesy Kendall / Hunt Publishing Co.

end of the cut consists of white, quartz-rich sandstone that contrasts sharply with the overlying beds. The face of the scarp is well-exposed right around the corner at the bottom of the hill by an old sand and gravel pit on the Pleasant Valley Road. The underlying Precambrian contact is not exposed, but the proximity of gneiss cuts suggests that it isn't far below the surface.

Between here and Alexandria Bay, all the roadcuts are either Precambrian gneisses or Potsdam sandstone. Near Alexandria Bay, the two units alternate in a manner that suggests the Potsdam beds have survived erosion only where they fill original pockets in the gneiss. In other words, the Precambrian topography, locally, is now much as it was at the time of invasion of the "Potsdam Sea," over 500 million years ago, despite its re-excavation as a result of the geologically recent uplift.

One of the most dramatic roadcuts in the North Country, the so-called Alex Bay cut, occurs near Cranberry Creek. It is easy to recognize because it is very large, with equally high cliffs on both sides. The downhill end of the cut exposes a knoll of Precambrian gneiss with steeply-inclined gneissic layering. Basal Potsdam beds partly drape over the knoll, and then pinch out against it, while the nearby, overlying beds simply thin out over the top of it. Upwards in the section, bedding becomes increasingly horizontal and uniformly thick, leaving little doubt that the irregularities of the Precambrian depositional surface are intact.

The contact displayed in the Alex Bay cut is an angular unconformity of profound dimensions. The rocks below are 1100 million years

294

old, and the Potsdam is more nearly 500 million years old; therefore, the contact represents a 600-million-year time gap! Such an uncomformity or time gap is a break in the rock record produced by either erosion or non-deposition. This one is an angular unconformity because the gneiss layers are upended and shaved off, forming a sharp angle with the overlying beds. Look closely at the gneisses just below the contact with the overhanging Potsdam beds and note the thin zone of deeply weathered and crumpled rock. Could this be a fossil soil zone?

Close-up of Precambrian-Cambrian unconformity in Alex Bay cut
Courtesy Kendall/ Hunt Publishing Co.

The Thousand Islands are the surface expression of the Frontenac Arch, and the visible connection between the Grenville Province of the vast Canadian Shield and the Adirondacks. The gneisses of the islands and Alexandria Bay area are mostly pink, massive rocks of granitic composition, called Alexandria Bay gneiss, or Thousand Islands gneiss, but darker, syenitic, and banded gneisses also occur. Note that most of the exposures along the highway are smoothly rounded, even polished by scouring ice. The islands themselves are elongated parallel to the river and display similar rounded form with pronounced asymmetry of cliffy, upstream ends and gentle, downstream trailing slopes. This indicates that glacial advance during the Ice Age was upvalley toward the southwest, more-or-less parallel to the river. These features may be viewed at close range by crossing the Thousand Island Bridge on Interstate 81 southwest of Alexandria Bay to Wellesley Island State Park. There, also, you will be able to see Potsdam sandstone exposed on the southwest flank of the arch, not the more familiar reddish-brown, but white.

Most of the landscape between the Thousand Islands Bridge and Clayton (7 miles) is flat tableland developed on Potsdam sandstone. About midway the Theresa forms a low scarp not far inland from the road. Potsdam ledges may be seen at the shore along much of this route.

Just southwest of Clayton, the topography is hillier where the road climbs over till hills. Black River limestone appears in several road-cuts between Clayton and Cape Vincent (15 miles). At Cape Vincent, the overlying Trenton limestones are well exposed along the shore, but there are not many roadcuts in bedrock of any kind.

Cape Vincent is at the point where Lake Ontario ends and the St. Lawrence River begins. The entire outflow from the Great Lakes passes here on its way to the Atlantic Ocean. The nearby opposite shore is not the Canadian mainland, but Wolfe Island, the largest of the Thousand Islands, completely capped by nearly flat-lying beds of Black River and Trenton limestones. The mainland is about 9 miles away, beyond the island.

The route between Cape Vincent and Watertown (24 miles) traverses tableland developed on Black River and Trenton limestones. The contact between these two units is near the road and zig-zags back and forth as it crosses a number of slightly entrenched streams, including Kents Creek, Three Mile Creek, the Chaumont River, Guffin Creek, and the Perch River, all of which drain southwestward into the lake. The Perch River, incidentally, goes underground at a place called Natural Bridge, about a mile downstream from Limerick, and then reappears a half mile farther on. The distant lake shoreline is deeply indented with bays as a result of drowning of the lower valleys of these streams. The Black River-Trenton contact invariably forms long Vs that point downstream, leaving little doubt that the beds dip, or slope, gently in that direction, away from the Frontenac Arch. In traversing downstream, you would thus pass from the older Black River to the younger Trenton beds.

Several good Trenton roadcuts begin at Fox Creek Road, 5.5 miles east of Cape Vincent. Small ledges of Black River limestone are visible in the hamlet of Three Mile Bay. About halfway, at Chaumont, you have a good view of the distant Tug Hill plateau—on clear days. Three miles east of there is an old quarry in Black River

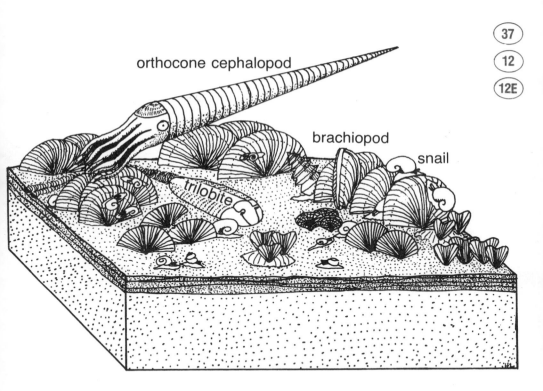

orthocone cephalopod

brachiopod

snail

trilobite

Artist's conception of a middle Ordovician life scene during Trenton (limestone) time. The orthocone cephalopod is similar to those found as fossils in the Chaumont quarry. Drawing by H. E. LaGarry

limestone. Exposed here are the upper beds of this unit which reveal beautiful cross-sections of large orthocone cephalopod shells, a tapering, straight, multicompartmented shellfish related to the modern Nautilus and octopus. The surface of the uppermost beds also displays glacial striae in two and, locally, three sets that record different passes of the ice. A miniature karst topography of caves and sinkholes is developed in the undisturbed beds behind the quarry, where joint fractures have been extensively widened by solution.

About one mile east of the quarry is the North Shore road to Pillar Point, a large promontory that separates Chaumont and Black River bays. The road loops around the point near the shore and comes back to 12E near Limerick, 17 miles from Cape Vincent and 7 from Watertown. Most of the shore is privately owned, but there are a few public beaches. There you can see step-like ledges of fossiliferous Trenton limestone that often show glacial polish and striae where soil has, until recently, protected them from deep weathering. Also common are large, rounded boulders of Precambrian gneisses that could only have arrived here by glacial transport from Canada.

At Brownville, note the old Brownville Hotel and other buildings constructed of light gray Black River limestone with peculiar thick seams of darker mortar. Between here and Watertown (3 miles), the road follows the Black River gorge with impressive limestone cliffs at its sides. The road here is built on top of this unit, and small erosional remnants of the overlying Trenton limestone can be seen here and there along the way.

Pillar Point, with glacially polished and striated bed of Trenton limestone, and glacial erratics of Precambrian gneiss
Courtesy Kendall / Hunt Publishing Co.

XI

THE ADIRONDACK
MOUNTAIN REGION

The Adirondacks are new mountains carved from very old rocks. Some of the original rocks may have formed as far back as three billion years ago, but all of those early rocks were intensely metamorphosed during the Grenville mountain-building event and thus carry a radiometric age of about 1.1 billion years. The newness of the topography results from the geologically much more recent uplift of the mountain mass. The rise continues today at the measured rate of about 3 millimeters per year! On the human clock, that small amount seems insignificant; but, on the geologic scale, it would raise the roof of the mountains a full three kilometers in the short span of a million years—less, of course, the altitude lost to erosion which generally proceeds more slowly on such hard rocks.

Most of the Grenville-age rocks of the Highlands Adirondacks are meta-igneous, metamorphosed anorthosite, granite, syenite, and a host of others. The metanorthosite occurs in several large bodies, in particular the enormous Marcy massif that underlies nearly all of the High Peaks region. These unusual rocks are of special interest because their peculiar mineral composition—90% or more plagioclase feldspar—requires extreme temperature and pressure conditions that exist only at enormous depths. In this case, a depth as great as 15 miles. The Marcy massif apparently intruded the country rocks, the remnants of which are now seen around it and locally intermingled with it, as a thick sill of magma. All were later extensively deformed and metamorphosed during the Grenville event. Much of the anorthosite was subjected to crushing, then subsequently healed by the metamorphism; still, relict crystals of plagioclase two feet long are occasionally found! Another reason for interest in these nearly monomineralic rocks is their close resemblance to some of the lunar rocks collected by the Apollo astronauts. The anorthosite is discussed in more detail in the Whiteface Mountain roadlog.

The northwestern, or Lowlands, Adirondacks, by contrast, are dominated by metasedimentary and metavolcanic rocks, with marbles and metasedimentary "paragneisses" most abundant.

The Grenville event was followed by an extremely long period—600 million years long, during which the ancestral Adirondacks were worn down to a surface of low relief. In late Cambrian time, a shallow inland sea covered all, or most, of the region. In this sea, a new rock record was built up on top of the old erosion surface in the form of flat-lying Cambrian and Ordovician strata. This blanket of sedimentary rocks probably survived at least until the post-Taconian block faulting shown by the many linear valleys of today's mountains; its nearly total removal occurred as a result of erosion associated with the more recent uplift. The uplift also tilted the strata away from the dome.

The Blue Mountain Lake region is of special interest to geologists because it is the center for extensive micro-earthquake activity and occasional larger, felt earthquakes. The micro-earthquakes cannot normally be felt, but they can be detected by sensitive instruments, and their records are used to study causes and prediction of larger, potentially destructive events. A swarm of hundreds of small earthquakes occurred in 1971 that enabled researchers from Lamont-Doherty Geological Observatory to make the first successful prediction of a larger earthquake in North America. The event, which took place on July 27, 1971, had a magnitude of only 2.5 (the 1964 "Good Friday" earthquake in Alaska had a magnitude of 8.6); but that was still significantly larger than the swarm earthquakes.

The cause of the earthquakes in this region is still unknown, despite the scientific attention they have received. The region lies between the well-defined, northeast-trending fault zones of Long Lake and Indian Lake. Numerous smaller faults have been charted near Blue Mountain Lake. Movement along any one of these faults would be capable of producing an earthquake of sufficient intensity to be felt. An unusually large event of magnitude 5.2 occurred on October 7, 1983, that was felt as far away as Philadelphia, Maine, Ottawa, and Buffalo. Lamont-Doherty seismologists, studying aftershocks from this earthquake found that movement occurred at depth along a fault which, when projected along its trend, would surface at Catlin Lake, 15 miles northeast of Blue Mountain Lake. Geologists are now searching the Catlin Lake area for any surface displacement that might be attributed to the fault movement.

A remotely possible cause of fault activity in this region is the contemporary doming of the Adirondacks. Blue Mountain Lake is near the center of the dome where the accumulated uplift, and stresses, are greatest.

The history of Wisconsin deglaciation in the Adirondacks is complex and difficult to envision without a series of maps showing where ice, water, and high ground were. In essence, however, the front of the wasting ice sheet retreated slowly northward. At first, as the ice thinned, it broke up over the Adirondacks and was gradually cleaved by the peaks and ridges. Valley glaciers were left on the higher mountains while many lower valleys were plugged by the ice sheet itself or remnants of it. The vast quantity of meltwater spilling from the glaciers during this period had no easy escape and was impounded between ice, bedrock slopes, and moraine dams, giving rise to temporary lakes, such as South Meadows Lake.

Many of these lakes persisted long enough at certain levels for normal shoreline features to develop—wave-cut benches, sea cliffs, beaches, deltas, and dunes. As ice dams melted away, new spillways opened at lower levels than the preceding ones; new lakes formed with different shorelines. Meanwhile, the valleys received flat-lying layer upon layer of sediments

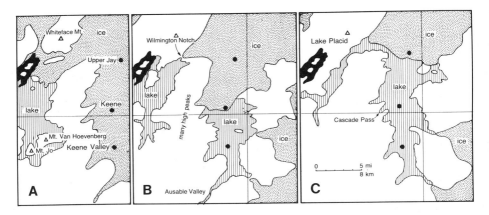

Three progressive proglacial lake stages in the Lake Placid-Keene Valley region; unblocking of Wilmington Notch permitted interconnection between two separate lakes. Lakeshore features can be seen now at various levels on the valley sides. Courtesy Kendall/ Hunt Publishing Co.

supplied in copious quantities from the melting ice and from erosion of the barren landscape. The deeply-scoured, northeast-trending fault valleys held finger-like extensions of these meltwater lakes. Many, like the Ausables, Cascades, Long, and Indian, remain imprisoned in their moraine-blocked valleys today. Others are now completely drained or filled up with sediments, and modern streams are cut into them in various degrees.

Nearly all the streams tributary to the St. Lawrence are entrenched to some degree. In general, the depth to which a stream can erode is controlled by a base level. When Lake Iroquois occupied the Ontario basin and upper St. Lawrence Valley, the stream base level was the lake level. Later, the draining of the lake combined with the rebound of the land surface lowered the base level, renewed the downcutting of the streams, and led to their current entrenchment.

South Meadows Lake plain and the High Peaks from North Elba; Avalanche and Indian passes are carved from northeast-trending fault zones.
Courtesy Kendall / Hunt Publishing Co.

NY 73:
Lake Placid—Underwood
33 mi./53 km.

This route is one of the most scenic and geologically interesting in the Adirondacks. East of Lake Placid, the road crosses the flat surface of the Lake Placid airfield and the fair grounds used for the opening and closing ceremonies of the 1980 Winter Olympics. Lake Placid lies in a broad valley surrounded by high mountains. By the fairgrounds, the Sentinel Range is seen as an impressive rampart bordering the valley on the east and northeast; scar-faced Whiteface Mountain lies almost directly north. Actually the valley floor was once much more level, as it was built upon flat-lying sediment layers deposited in South Meadows Lake, a meltwater lake formed during the closing stages of Wisconsin glaciation. Downcutting by modern streams is responsible for much of the relief you see on the valley floor today. The presence of the original valley, however, is probably due to the lesser resistance to erosion of the rocks beneath it relative to the hard anorthosite of the enclosing mountains.

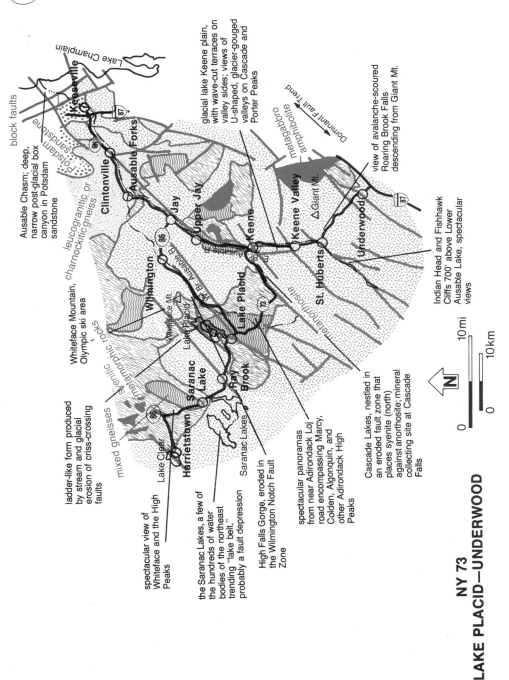

block faults

Lake Champlain

Keeseville

Potsdam sandstone

leucogranitic or charnockitic gneiss

Ausable Chasm; deep, narrow post-glacial box canyon in Potsdam sandstone

Clintonville

Ausable Forks

90

87

Jay

Upper Jay

E. Br. Ausable R.

glacial lake Keene plain, with wave-cut terraces on valley sides; views of U-shaped, glacier-gouged valleys on Cascade and Porter Peaks

metagabbro amphibolite

Dominant Fault Trend

Keene Valley

△Giant Mt.

view of avalanche-scoured Roaring Brook Falls descending from Giant Mt.

Whiteface Mountain, Olympic ski area

86

Wilmington

W. Br. Ausable R.

Whiteface Mt.

Lake Placid

Keene

73

St. Huberts

Underwood

87

syenitic rocks

metamorphic rocks

Saranac Lake

Ray Brook

metamorphosite

Harrietstown

86

Lake Clear

Saranac Lakes

mixed gneisses

Indian Head and Fishhawk Cliffs 700' above Lower Ausable Lake, spectacular views

ladder-like form produced by stream and glacial erosion of criss-crossing faults

spectacular view of Whiteface and the High Peaks

the Saranac Lakes, a few of the hundreds of water bodies of the northeast trending "lake belt," probably a fault depression

High Falls Gorge, eroded in the Wilmington Notch Fault Zone

spectacular panoramas from near Adirondack Loj road encompassing Marcy, Colden, Algonquin, and other Adirondack High Peaks

Cascade Lakes, nestled in an eroded fault zone that places syenite (north) against anorthosite; mineral collecting site at Cascade Falls

N

0   10 mi

0   10 km

NY 73
LAKE PLACID—UNDERWOOD

304

A short distance east of the fairgrounds, the road descends to the west branch Ausable River past a sand and gravel quarry and the Olympic ski jumps. The quarry material presumably represents some of the valley fill. The road then passes back up onto another segment of the dissected lake plain with excellent open views to the south near the Adirondac Loj road (3.5 miles from Lake Placid). The highest, rather bald summit visible in this Montana-like view is Mt. Algonquin, the Empire State's second highest peak, at 5112 feet above sea level. The deep cleft to the right or west of the peak is Indian Pass, by the 2000-foot cliff of Wallface Mountain. The pass follows one of the many northeast-trending fracture zones, where the rock is crushed and easily eroded. The first high peak east of Algonquin is Mt. Colden, and Avalanche Pass by its side is another one of these excavated crush zones. The fracturing and crushing in most of these zones dates back approximately 400 million years to the Taconian

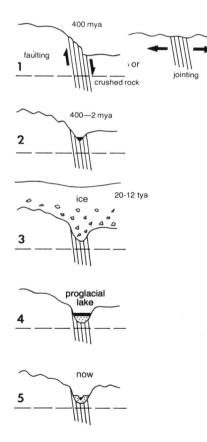

Evolution of linear valleys in the Adirondacks, especially those with a northeasterly trend. Note that only one of four glaciations is shown, and intermittent stream erosion is omitted.

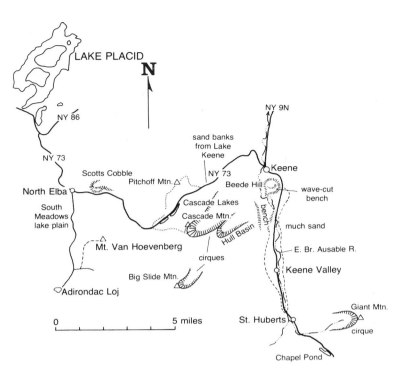

Some of the glacial features visible along NY 73

mountain building event centered principally in adjoining New England. Presumably, the fault zones were already carved out by streams to form deep mountain valleys when the first of four Pleistocene glaciations began about two million years ago. Their northeasterly alignment, nearly parallel to glacial advance, facilitated gouging by ice into the U-shaped cross-profile typical of glaciated valleys in mountains everywhere.

Valleys that lay athwart the direction of ice advance, meanwhile, received less erosion, and more glacial debris was dumped into them. As in other parts of New York State, however, this glacial history of the Adirondacks can only be well-documented for the last, or Wisconsin, glacial stage.

Aside from the flat-floors of the valleys, the most conspicuous glacial lake features you will see from the road are the wave-cut benches at various levels on the valley sides, but they require a trained eye. There is a prominent one on Scott's Cobble, a knobby hill on the north side of the road about one mile east of the Loj road. The shelf is camouflaged by a thick stand of pine, which grows well in the sandy soil. You may be able to pick it out as you go by. This one, at 2209 feet above sea level, marks the highest elevation reached by any of the meltwater lakes in the Adirondacks.

Three miles east of the Adirondack Loj road is the entrance to Mt. Van Hoevenberg winter recreation center, where the cross-country, biathlon, bobsled, and luge competitions were held during the 1980 Winter Olympics. From here, you have a clear view to the northeast of Cascade Pass, with the bare rocky rib of Pitchoff Mountain on the left, and the steep, wooded slopes of Cascade Mountain on the right.

Narrow Cascade Pass, with its two lovely, slim lakes, is 8 miles east of Lake Placid. A good stopping place is at the picnic area between the upper, or west, and the lower lakes. The pass is carved along yet another of the northeast-trending fracture zones, this one extending for at least ten miles, from near Adirondack Loj to Keene. The rock of the Cascade Mountain side is anorthosite, and the Pitchoff side, syenite.

Cascade Brook, on the slopes near the picnic area, is a famous mineral collecting locality. The source of the many different minerals is above the falls, but this is difficult to reach and dangerous, with highly fractured rock and landslide debris. The collecting in the streambed below the falls is excellent. Look for stream-washed spec-

Cascade Pass and Lower Cascade Lake, another northeast-trending fault zone; rock on right is anorthosite, and on left is syenite
Courtesy Kendall / Hunt Publishing Co.

View from East Hill of Cascade and Porter mountains, the glacial cirque between, and the Hull Basin cirque to the left of it
Courtesy Kendall/ Hunt Publishing Co.

imens of blue calcite, heavy black magnetite, coarse-grained, whitish marble with tiny green diopside crystals, black hornblende with shiny cleavage surfaces, glittering brownish sphene, iridescent plagioclase feldspar that shows a play of colors on wetted surfaces, reddish garnet, and a rare mineral called monticellite, that forms tiny, dull, tannish grains in marble.

Between Cascade Lakes and Keene (5 miles), the road descends quickly about 1000 feet between steep slopes littered with large talus blocks. Halfway, there is a roadcut in clean sand that was opened during road construction in 1979, but is now grassed over. This sand is a high shoreline deposit of Lake Keene, another meltwater lake that filled Keene Valley at the time of the South Meadows Lake mentioned above. In Keene, the road crosses the east branch Ausable River. The stream here makes a detour around the village through a chaotic mixture of gneisses and marbles that may be part of a large "roof pendant" that projected into the original mass of anorthosite magma as it intruded into the existing country rock. Alternatively, it might be a large xenolith, or block of country rock that broke off and fell into the magma.

Between Keene and Keene Valley (4 miles), the road traverses the level Lake Keene plain most of the way, past the landing strip at

Marcy Field. A good place to view the mountains and valley is by the cemetery near where 9N branches off to Elizabethtown. From here, you will be able to survey the Keene Valley sides for traces of wave-cut benches, which are there, but hard to see. One well-developed one is on Beede Hill just to the north of the cemetery, at about 1250 feet above sea level.

In Keene Valley village, the road passes over Johns Brook, just before it joins the Ausable River. The bed of the stream is positively choked with well-rounded cobbles undoubtedly washed out of the valley fill.

Most of the way between Keene and St. Huberts, the road follows the east branch Ausable River, here a very clear, rocky stream. At St. Huberts, the highway crosses over the river where it issues from the Ausable Valley, perhaps the most majestic of all the Adirondack valleys. The land is presently owned by the Ausable Club, which succeeds the original Adirondack Mountain Reserve. The Reserve was formed by a group of sportsmen in the 1800s to buy and restore the land ravaged by lumbering and forest fires. Today, we owe a great deal to their efforts, for the slopes are once more verdant, and the deep blue waters of the Ausable Lakes are clear and clean. The Reserve holdings were once much larger, but more than 18,000 acres of restored land were sold to the state in the 1920s, including Mt. Marcy.

Like so many Adirondack valleys, the evolution of the Ausable Valley is closely tied to Wisconsin glaciation. In the early stages of glacial retreat, the Ausable Valley tongue of the Keene Valley lobe accomplished the final gouging of the bedrock floor and walls. Meanwhile, other ice tongues were doing the same to the Wilmington Notch, Cascade Pass, and Johns Brook fracture zones. Later, the valley became the outlet for Lake Keene, which was blocked on the north by the ice itself. As the ice receded yet farther northward and new outlets were opened, the level of the lake dropped, so that the Ausable spillway was abandoned. Water remained, however, as a single lake dammed at either end by glacial moraines. Today, the original lake is divided into the upper and lower Ausable lakes by the alluvial fan of Shanty Brook, with its headwaters on the high peaks to the west.

The two best places to enjoy the stunning beauty of this valley are at the golf course in front of the clubhouse and at Indian Head above Lower Ausable Lake. The first may be reached just by driving up the club road from St. Huberts. From this open vantage point, the massive, slide-scarred summit of Giant Mountain (4627 feet) looms incredibly large and close, appearing to hang over your head. The second may be reached by a three-mile hike to Lower Ausable Lake, and then a steep half-mile climb. The cliffs are above a bend in the northeastern end of the lake. Near the summit, the trail descends over several step-like ledges. The final one forms a broad terrace that appears, from above, to be floating over the lake, and the effect is breathtaking. It is as if you could step off the edge and drop 700 feet straight down into the water.

Lower Ausable Lake looks more like a fjord than anything else. The steep, green-carpeted sidewalls descend beneath the Persian blue waters of the lake without a break in slope, prohibiting trails on either side. Upper Ausable Lake shimmers in the distance, set in a primitive, forested landscape of low, rounded hills, a scene of natural splendor unbroken by any visible habitation. To the northwest, you can look right into the awesome bare rock cirque of Gothics, and in the opposite direction gaze up to the summits of Colvin, Dial, and Nippletop.

On the hill between St. Huberts and Chapel Pond, the road passes

The Ausable Lakes from Indian Head (Marcy-type anorthosite in foreground)

the bare cliff of Roaring Brook Falls on the east across a little valley. This harmless-appearing little creek, which drains the large western cirque of Giant Mountain, was the site of massive landslides and muddy torrents on June 25, 1963. The freshly exposed rock of the cirque and valley floor above the falls has been called a geologist's paradise because of the many details of anorthosite geology it reveals. The short hike to the lip of the falls begins from a small parking area near the base of the hill. This is also the starting point for trailless climbs of Giant up the bare walls of the cirque.

At the top of the hill, Chapel Pond is nestled snuggly between Round Mountain on the west and Giant Mountain on the east. This is one of the most favored rock climbing areas in the mountains, on the cliffs behind the pond, and on those around Giant Washbowl. The latter can be easily reached on the trail beginning just south of the parking area by the pond.

Between Chapel Pond and the US 9 junction, the road follows the north fork Bouquet River past numerous cuts in anorthosite with its distinctive, light-greenish weathered color. Nearly all of the cuts between here and Interstate 87 junction display prominent northeast-trending fractures at right angles to the road. A study of the topographic and geologic map suggests that these are genetically related to several more large-scale fracture zones parallel to Ausable Valley and Cascade Pass.

Whiteface Mountain and the Olympic ski trails

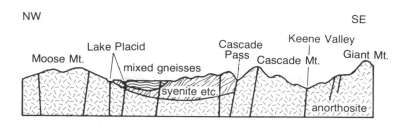

Schematic NW-SE cross-section through Lake Placid-Keene Valley showing numerous faults and rock types that control topography.

See map page 304.

NY 86:
Lake Clear Junction—
Lake Placid—Jay
36 mi./58 km.

Between Lake Clear Junction, at the NY 30 junction, and Harrietstown (4 miles), the road traverses a rather level lake plain past the Adirondack Airport at an elevation of slightly over 1600 feet. This plain is part of the "Lake Belt" of the Adirondacks, which, in addition to Lake Clear, includes the Tupper Lakes, Saranacs, and St. Regis Lakes. There are, in fact, over 50 lakes a mile or more in diameter, and well over a hundred smaller ones. Geologically and topographically, the lakes lie in an anomalous feature of low elevation and relief called the Saranac intramontane basin, which stretches for 35 miles in a northeasterly direction and varies from 10-15 miles wide. High mountains line the southeastern side and lower foothills line the northwestern side. Its elongation, parallel to many Adirondack fracture zones, suggests that it originated as a down-dropped block of crust bounded by faults. The few bedrock hills, principally of anorthosite, jut only 460 feet or less from the sediment fill. Two of the rock knobs guard the junction with NY 192A at Harrietstown. For further discussion of the glacial meltwater lakes of the Adirondacks, see NY 73 from Lake Placid—Underwood.

The panorama of the High Peaks from Harrietstown is one of the best in the Adirondacks, especially from points on 192A just north of the 86 junction. The pointed crest of Whiteface pierces the sky to the east, and many of the remaining High Peaks form the sawtoothed skyline to the south and southeast. In any of these directions, the lower intervening hills are lined up like giant steps to the heavens, an effect that is enhanced by hazy visibility.

Between Harrietstown and Saranac Lake village (5 miles), visibility is mostly restricted by trees, except by Lake Colby, 2 miles north of the village. The few roadcuts expose gneisses not far from contact with the enormous mass of Adirondack anorthosite that underlies nearly all of the High Peaks. The contact is crossed on the east side of the village. The Saranac Lakes, including Upper, Middle, and Lower, and numerous smaller lakes, lie generally west and south of the village.

Visibility is also restricted by trees over most of the 8 miles between Saranac Lake village and Lake Placid. Best views of the mountains are by the golf course on the east side of Ray Brook and coming down the final hill into Lake Placid. Ray Brook is the site of the Olympic Village used to house athletes during the 1980 Winter Olympics at Lake Placid, and since converted to a prison.

The distinctive slide-striped, pointed mass of Whiteface Mountain hovers over the village of Lake Placid. The peak is actually about 7 miles distant, but the lake is in between, and the lack of intervening mountains to obstruct the view makes it appear bigger and closer. The slide stripes, so common on the high Adirondacks are an interest-

Slide-striped northwest face of Mt. Colden and Avalanche Lake
Courtesy Kendall / Hunt Publishing Co.

ing geologic phenomenon. Nearly all of them occur on the smooth anorthosite bedrock, which has been carved and polished into steep-sided domes by stream and glacial erosion of criss-crossing fracture zones (see discussion of NY 73). Summit regions have been further steepened by alpine glaciers that coursed down their sides in the closing stages of the Ice Age. Had these persisted long enough, they would have carved Matterhorn-like peaks out of the domes. Because the slopes are steep and only sparsely fractured, the thin soil that eventually develops has only a precarious hold. Excessive wetting during spring thaw or heavy rains causes frequent landslides that peel soil, brush, and trees from the rock as if it were a giant band-aid strip. Whiteface has been more deeply scalloped by alpine glaciers than most of the High Peaks, yet has only a small number of slide scars. Most of them are in the large glacial cirque of the east side, near the Olympic ski slopes.

Lake Placid village is atop a hilly moraine that dams up the southwest end of Lake Placid. Most of the business district, however, is on Main Street by Mirror Lake, a much smaller lake of probable kettle origin. The peculiar ladder-shape of Lake Placid is eroded in cross-crossing fracture zones. Before glaciation, these were carved out by streams. They were later further hollowed out by southward advance of the continental ice sheets into and over them. The long arms were more deeply gouged and widened than the "rungs" because they were more nearly parallel to ice movement. During deglaciation, the ice front paused long enough near the village site to dump great mounds of debris that effectively blocked drainage to the southwest. The lake now has a small outlet through the moraine dam that passes around the west side of the village to the Chubb River, which, in turn, quickly joins the west branch Ausable.

Between the NY 73 junction at the south edge of Lake Placid village and Wilmington (13 miles) is some of the most spectacular scenery of the Adirondacks. There are open views of the High Peaks to south and southeast by the Lake Placid golf course. From two and a half miles east of the village, to the Flume, the road follows the west branch Ausable River. This is a clear, bouldery mountain stream, with bedrock banks, frequent rapids, waterfalls, and quiet pools that attract many trout fishermen in season. On the way, you pass through the narrow canyon called Wilmington Notch, with high cliffs on both sides. With little room to spare, the road winds back and forth practically on top of the river. This is the site of High Falls Gorge, a commercial tourist attraction where, for a fee, you can walk down into the gorge along a system of trails, stairs, bridges, and walkways anchored to the walls. Here, the Ausable River drops more than 100 feet in several lovely falls over a resistant mass of granitic rock. The

High Falls Gorge of the west branch Ausable River, carved from Wilmington Notch fault zone in granite   Courtesy Kendall/ Hunt Publishing Co.

reason for Wilmington Notch is revealed by features in the bedrock of the gorge; it is broken by numerous fractures that dip steeply to one side, so that downcutting by the stream leaves that side overhanging. This zone is just one of many northeast-trending fracture zones of the

Adirondacks that have been preferentially excavated by stream and glacier (discussed further in NY 73: Lake Placid—Underwood). The Wilmington Notch fracture zone extends at least from Lake Placid village to Wilmington. Close examination from the first observation deck by the upper falls (down a long flight of stairs) shows that the overhung side is highly brecciated (broken-up) in addition to being cut by parallel joints, which further facilitates erosion. Numerous dark gray basaltic dikes have been intruded along the joints. The uppermost falls drop through a slot where one of these has been worn away. Several generations of basaltic dikes have been recognized in the Adirondack region intruding the Precambrian rocks, the youngest being only 135 million years old. Those of High Falls intrude fractures presumably resulting from the Taconian mountain-building event, which would place their age at somewhat less than 400 million years.

The gorge is a great place to see potholes. These are rounded and polished hollows carved by the swirling of sand and gravel against bedrock in stream eddies. Many of those you see high above the river now are really still active, being carved deeper each spring when the water may rise 10 feet or more.

Other places to see the fracture zone close-at-hand, and free, are at the Whiteface Ski Center nearby, or at the Flume, 2 miles farther toward Wilmington. Views of Whiteface are especially good on this stretch of road.

At Wilmington, route 86 continues east to Jay at a cross-road, whereas the road to the left is the Whiteface Mountain Memorial Highway covered in the following chapter. Most of the way to Jay (3.5 miles), the views are obscured by trees. Two outstanding exceptions are at Beaver Creek meadow, the lowest point, and at the ski slopes at the highest point. From each place, Whiteface is seen in its true stature, as a lone monarch set apart from the other High Peaks, rising a full 4000 feet above the meadow, a great mass of hard rock triumphant over millions of years of unending onslaught by weather and erosion.

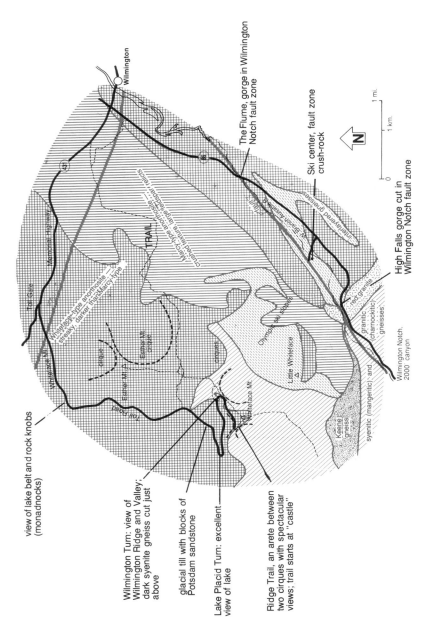

WHITEFACE MOUNTAIN MEMORIAL HIGHWAY

view of lake belt and rock knobs (monadnocks)

Wilmington Turn: view of Wilmington Ridge and Valley; dark syenite gneiss cut just above

glacial till with blocks of Potsdam sandstone

Lake Placid Turn: excellent view of lake

Ridge Trail, an arete between two cirques with spectacular views; trail starts at "castle"

Wilmington

The Flume, gorge in Wilmington Notch fault zone

Ski center, fault zone crush-rock

High Falls gorge cut in Wilmington Notch fault zone

Toll Gate

Memorial Highway

TRAIL

Whiteface-type anorthosite — streaky, darker fine-Marcy-type

Marcy-type anorthosite; large feldspar; relics

crushed anorthosite

Whiteface Mt.

cirque

Esther Mt. cirque

Esther Mt.

cirques

Whiteface Mt.

Little Whiteface

Olympic ski slopes

"Bierut" syenite

limey

interlayered gneisses

red granite

granitic (charnockitic) gneisses

syenitic (mangeritic) and

Keene gneiss

Wilmington Notch, 2000' canyon

Toll Road

N

0   1 mi.

0   1 km.

Whiteface Mountain
Memorial Highway
11 mi./18 km.

Whiteface is the only one of the major Adirondack peaks with a road up it. This is the highest spot to which a car may be driven in the state, and it has become increasingly popular as an easy way to survey some of New York's most breathtaking scenery. More than 40,000 cars make the ascent in a typical season. The road was built in the Depression years of the 1930s, to honor New York's war dead.

Beginning at Wilmington, the road climbs steeply up the first three miles to the toll gate, rising more than 1200 feet. In the lower part of the climb, you have several fleeting glimpses up to the left into the 1000-foot headwall of the Esther Mountain cirque. This mountain is really a northern buttress, or shoulder, of Whiteface. At two and a half miles, you pass the entrance to the Atmospheric Sciences Research Center of the State University of New York at Albany, a good place to look out over the west branch Ausable Valley and to get an even closer view of the Esther Mountain cirque. The true spectacle of Whiteface begins beyond the tollgate, on the eight and a half mile climb to the end of the road. The toll at this writing is $3 per adult, a rather steep fee for a carload of people, but, in my opinion, the scenery is worth it.

The road first angles up the north and west sides of Esther Mountain with sweeping views of the Saranac intramontane basin. Then it climbs up the Whiteface summit pyramid across its furrowed, avalanche-scarred, north flank, with numerous cuts in the pale gray anorthosite and in bouldery, sandy, glacial drift.

319

*Glacial drift high on
Whiteface Mountain
containing cobbles of
Potsdam sandstone*
Courtesy Kendall / Hunt
Publishing Co.

A good place to stop and look closely at drift is at about 3 miles, by a parking area. The presence of drift, here, at nearly 4000 feet above sea level, shows positively that the ice was at least this high up on the mountain. There is ample evidence, in fact, that it overtopped the Adirondacks by as much as a mile during the climax of Wisconsin glaciation, about 22,000 years ago. Found in this drift bank are suspected fragments of Potsdam sandstone. The nearest bedrock outcrops of this unit lie at an elevation of 1000 feet, 17 miles to the northeast. Those nearest to the north are 30 miles away at 1500 feet above sea level. This would mean, of course, that pieces of Potsdam sandstone were plucked by the southward-advancing glacier, lifted to 4000 feet or more as the ice rode over the mountain, and then dropped here as it later melted. This is not surprising when you consider that the continental ice sheets at climax had spread from Labrador to Long Island, a distance of about 1500 miles!

A short way beyond the drift bank, the road makes the first of two hairpin turns atop the sharpest arete on the mountain. Aretes form where the headwalls of two opposing alpine, or valley, glaciers meet; this one separates the north and west cirques. The splendid view of the lake from here gives this place its name of Lake Placid Turn and makes it a worthwhile stop.

Anorthosite is a rock that contains 90% or more of plagioclase feldspar, with most of the balance consisting of pyroxene. With increasing pyroxene content, anorthosite grades into gabbro, which is much darker-colored and contains much more iron and magnesium. A very large mass of anorthosite called the Marcy massif underlies all of the High Peaks region of the Adirondacks. According to current

320

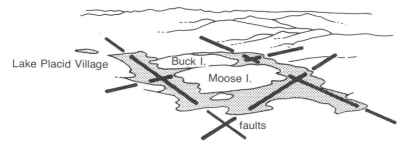

*Lake Placid view from the Lake Placid turn, showing the ladder-like shape
and controlling faults*

theory, this large body originated as an igneous mass of partially or
wholly melted rock that forcibly intruded the surrounding rocks at
great depth, and along with all the other Adirondack rocks, was later
metamorphosed during Grenville mountain-building. Strictly speaking, therefore, the rock should be called metanorthosite, since it has
been metamorphosed.

The rock on the inside of the curve is Whiteface-type anorthosite,
with fairly abundant red garnet, and lenses, blotches, and streaks of
black pyroxene. The other principal anorthosite variety in the Marcy
massif, called the Marcy-type, is generally much coarser, less
streaky, and contains partly crushed grains of grayish, greenish, or
bluish plagioclase that occasionally reach a foot or more in length,
but which were larger before crushing. These are set in a whitish, or
pale greenish matrix composed of finely crushed plagioclase welded

Marcy-type anorthosite (A) and Whiteface-type (B); scale bar is 2 cm. long.
Courtesy Kendall/ Hunt Publishing Co.

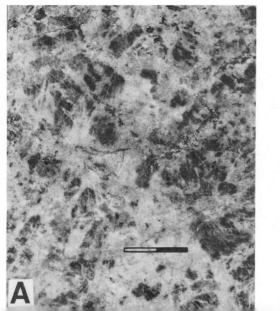

together by the metamorphism. The Whiteface-type is a border facies of the massif near its contact with the surrounding gneisses; most of the mass is Marcy-type. Note that the retaining wall at the Lake Placid Turn contains several representative rock types of the region, including both types of anorthosite; charnokite—a streaky granitic rock containing a variety of dark pyroxene; and amphibolite—a nearly black rock made up of dark, shiny hornblende and plagioclase in about equal proportions. Several of the rocks contain abundant red garnet.

The next hairpin turn, called the Wilmington Turn, overlooks Esther Mountain and the Wilmington Range, extending northeastward or to right as you look outward from the lower end of the turn, alongside the Wilmington Valley. Here, you also have a good view to the rim of Esther Mountain cirque, a half-mile away across a rather flat saddle. The hiking trail to the summit of Whiteface comes up the near side of the cirque, then straight to here, where it emerges briefly before continuing straight to the peak along the ridge.

At this point you cross the boundary from anorthosite to syenite, the dark rocks of the long deep cut beyond the turn. Well before the parking area at the end of the highway, however, you pass back into the anorthosite. From the parking area, you may reach the summit either by elevator or by climbing the ridge on the uphill side of the "Castle." Many people elect to take the elevator up and then walk

Ridge trail from the "Castle" to the summit of Whiteface Mountain, showing steep cirque headwall, slide stripes, and sheet-jointing of bedrock (near-level fracturing)   Courtesy Kendall/Hunt Publishing Co.

down. The rise is equivalent to 26 stories. The ridge trail follows the same glacially-sculptured arete as that of the Lake Placid Turn. The view along the way is positively breathtaking, so be sure to follow the trail at least one way, up or down. On the way, note the closely-spaced fractures visible in the vertical headwall of the cirque that slope, or dip, gently to left as you go up. These are called sheeting, a special kind of fracturing that results when once deeply buried rocks expand as their overburden is removed by erosion. These rocks are believed to have formed at depths as great as 15 miles!

Because Whiteface stands alone far from the other High Peaks, the view from the top is especially good. On clear days, you can follow the tiny ribbon of the St. Lawrence River all the way to Montreal, and looking northeastward, beyond Wilmington Valley, you can see the shimmering waters of Lake Champlain. In both directions you will look down the dissected slope of the Adirondack Dome to the flat lowlands country. Surveying the region generally to the south, it is possible to pick out the distinctive profiles of many of the High Peaks. Looking down beyond the top of the chairlift on Little Whiteface Mountain to the south, you can peer right into the shadowy abyss of Wilmington Notch. This view leaves little doubt of the important role played by the Notch in controlling drainage during the meltwater lake stage of Wisconsin deglaciation, as the Ice Age drew to a close (see NY 73 discussion).

"Demoiselles" in glacial outwash. The pebble "umbrellas" shield their sandy pedestals from washing away in a rainstorm. They are about one inch high.

Ausable Chasm, cut in Keeseville member of the Potsdam sandstone

See map page 304.

NY 9N:
Ausable Chasm—Keene
30 mi./49 km.

Ausable Chasm is located 1 mile north of Keeseville on US 9. The land at this writing is owned by the Ausable Chasm Company, which charges admission to enter the canyon and also operates a boat ride through the rapids section.

Ausable Chasm is a deep box canyon of the Ausable River, cut into a thick, nearly flat-lying section of the Keeseville member of the Potsdam sandstone. It owes its origin to events of Wisconsin glaciation. During much of that glacial episode, the Champlain ice lobe pushed southward along the Champlain Valley and ground away at

325

the bedrock at its sides. This produced cliffs here and in numerous other places that were later mantled with glacial debris as the ice retreated. During glacial recession the valley was first occupied by glacial Lake Vermont when the St. Lawrence Valley was still blocked by ice, and later by the saline waters of the Champlain Sea when the St. Lawrence was finally opened to the Atlantic. Water levels then were much higher than those of the modern Lake Champlain, and rivers like the Ausable, which soon established themselves above the valley sides, were only able to cut down into bedrock to those levels. In other words, their base levels of erosion were established by the height of the lake surface. Since then, the base level has been lowered significantly by the rebound of the land surface that had been depressed under the weight of the ice. This initiated canyon development at the water's edge where the river spilled over the glacial debris. The canyon quickly extended itself upstream through the unconsolidated sediments to the buried Potsdam scarp, where falls formed. The chasm has since been carved by the much slower recession of the falls through the hard bedrock.

The chasm follows prominent vertical jointing. Erosion at the falls causes block after block of joint- and bedding-bounded sandstone to break off and join the sediment of the streambed. The river follows the strongest joint set but there are numerous small side canyons excavated along weaker cross-joints. The chasm remains so deep and narrow because the horizontal stratification of the beds is very stable, like stacked books, and also because the quartz-rich sandstone resists weathering. Once the river has done its work of downcutting, the

Strong vertical jointing near upper end of Ausable Chasm; bedding is horizontal; Potsdam sandstone. Courtesy Kendall/Hunt Publishing Co.

joint-bounded walls remain intact for a very long time. Modern reces- sion of the falls is effectively halted by construction of a dam at Rainbow Falls, and diversion of much of the flow for generating electricity.

Downstream from the chasm, the Ausable River is cutting deeply into its own delta, which was formed in Lake Vermont and the Champlain Sea. Those original deltaic sediments are, meanwhile, being redistributed in the form of a new delta that extends farther out into the modern Lake Champlain at a lower elevation. Upstream from Ausable Chasm, in Keeseville, the upper beds of the Potsdam sandstone section are exposed in the streambed and banks. Between Keeseville and Clintonville (6 miles), you pass over the contact with the anorthosite and a granitic gneiss like that of the great cliff of Poke-O-Moonshine Mountain to the south. There are several open views of the mountains en route, including a good one of Whiteface Mountain at Clintonville. The valley floor up to and beyond Clinton- ville is quite flat. This is the delta-level that extended inland from the shores of Lake Vermont, drowning the valley of the Ausable River. As the continental ice sheet retreated slowly northward, meltwater filled valleys like this. This resulted in not only deltas built by inflowing streams, but also beaches, wave-cut benches, sea cliffs, and coastal dune formation. Note the many sandy, gravelly cuts along this stretch of road, as well as bedrock.

The delta plain ends rather suddenly two and a half miles west of Clintonville, as the road enters a V-shaped valley that continues to beyond Ausable Forks (5 miles). The transition occurs at about 530 feet above sea level, or about 30 feet above the delta plain at Clinton- ville. Ausable Forks is so-named because it lies at the junction of the two branches of the river. The northern bridge in the village crosses the west branch, and the southern one, which the road passes but doesn't cross, goes over the east branch. The west branch originates near Lake Placid, and flows by Whiteface Mountain and Wilmington before reaching here (see NY 86 discussion). The second starts in the Ausable Valley and flows through Keene Valley to here (see NY 73 discussion). The east and main branches, in particular, have been the scene of disastrous ice jams in years when the spring thaw came too suddenly.

Between Ausable Forks and Keene (16 miles) is one of the most peaceful, grassy valleys in the Adirondacks. Visibility is open, with frequent glimpses of high mountains east and west. Roadcuts along the way are either in white anorthosite, streaky gneiss, or in the abundant sand and gravel that was dumped in the valley during the glacial Lake Keene stage. South of Ausable Forks is the entrance to one of two quarries where Adirondack anorthosite is taken for dimension stone and monuments. The two quarries yield two principal varieties of stone, the "Lake Placid Blue" and the "Cold Spring Green." Quarrying of the anorthosite is accomplished by the painstaking technique of wire-sawing, with a twisted steel wire. Using a complex array of pulleys, a great length of this wire first passes through a slurry of grinding compound, which lodges in the grooves. It then passes over the rock, which is slowly cut by the grinding compound, not by the wire. Slabs thus removed are later trimmed, engraved, and polished in the mill at the quarry. The location of the Ausable Valley, from Keeseville at least to Upper Jay, appears to be controlled, like so many other linear valleys of the Adirondacks, by a northeast-trending fracture zone (see NY 73 and 86 discussion).

Lake Placid Granite Company quarry in the anorthosite at Ausable Forks. Most of the cutting is done with wire saws.
Courtesy Kendall/Hunt Publishing Co.

"Herkimer diamonds" from Tasselsville, N.Y. These are exceptionally clear, well-formed crystals of quartz for which the Little Falls dolostone is famous.

NY 28:
Herkimer—Blue Mountain Lake
103 mi./167km.

Herkimer is on a broad, flat sediment plain of the Mohawk River where it is joined by West Canada Creek from the north. Between Herkimer and Middleville (10 miles), you follow the West Canada Creek valley, carved in Cambrian and middle Ordovician sedimentary rocks. The strata are gently uptilted against the Precambrian rocks of the Adirondack dome exposed a few miles to the northeast. The plateau-like uplands on either side of the valley are capped by middle Ordovician Utica shales. The terraces in places on the valley sides are bedrock platforms on middle Ordovician Trenton/Black River limestone beds. The floor of the valley is in the Cambrian Little Falls dolostone, which is exposed in several roadcuts. This unit is famous because it contains the so-called "Herkimer diamonds." They're not really diamonds, but small, water-clear quartz crystals found in solution pockets or "vugs" of the dolostone. Occasionally larger crystals, four or more inches in diameter, are also found, but they are usually less clear, imperfectly formed, and contain many inclusions. Middleville has, in the past, been the center for collecting.

The route continues in the West Canada Creek valley between

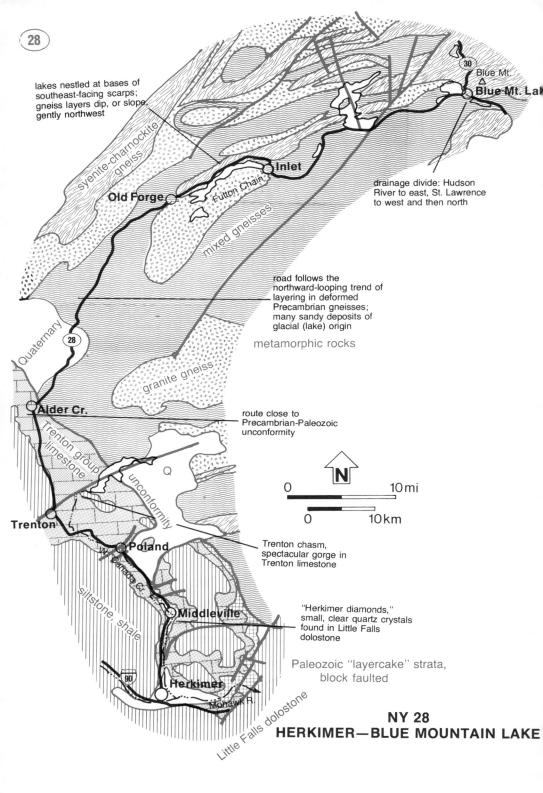

lakes nestled at bases of
southeast-facing scarps;
gneiss layers dip, or slope,
gently northwest

syenite-charnockite gneiss

Inlet

Fulton Chain

Old Forge

mixed gneisses

drainage divide: Hudson
River to east, St. Lawrence
to west and then north

30

Blue Mt.

Blue Mt. La

road follows the
northward-looping trend of
layering in deformed
Precambrian gneisses;
many sandy deposits of
glacial (lake) origin

metamorphic rocks

Quaternary

28

granite gneiss

Alder Cr.

Trenton group limestone

route close to
Precambrian-Paleozoic
unconformity

Q

unconformity

N

0   10 mi

0   10 km

Trenton

Poland

W. Canada Cr.

Trenton chasm,
spectacular gorge in
Trenton limestone

siltstone, shale

Middleville

"Herkimer diamonds,"
small, clear quartz crystals
found in Little Falls
dolostone

90

Herkimer

Mohawk R.

Little Falls dolostone

Paleozoic "layercake" strata,
block faulted

NY 28
HERKIMER—BLUE MOUNTAIN LAKE

Middleville and Trenton (17 miles), still skirting the Precambrian boundary. En route you pass the village of Poland which, strangely, is in the Herkimer town of Russia! Upstream from Trenton, West Canada Creek has been dammed to form the Hinckley Reservoir. A section of the valley below the dam near the hamlet of Trenton Falls is a spectacular gorge called the Trenton Chasm, cut in the Trenton limestone where the river formerly dropped about 300 feet in 2.5 miles over several waterfalls. Unfortunately the chasm is now developed by Niagara Mohawk Power Corporation for generation of hydroelectricity, and the public is forbidden to trespass. You may, however, view the lower end of the gorge from a bridge at Trenton Falls, and the upper end from the bridge at Prospect, about 2 miles north of Trenton Falls.

Between Trenton and Alder Creek (12 miles), the road follows Cincinnati Creek over fairly level upland in the Trenton limestone. The landscape changes to heavily-wooded low mountains beginning just northeast of Alder Creek, as the highway enters the Precambrian terrane.

It would be nearly futile to try to point out each and every rock type in roadcuts between Alder Creek and Blue Mountain Lake (64 miles), and try to fit each into some kind of meaningful geological pattern. Roadcuts along the route are, in the first place, unusually few and far between, and secondly, the large-scale regional patterns indicated by outcrops of the various metamorphic rock units are extremely complex. Extensive study of the southern Adirondacks indicates the rocks there have been folded in different directions at least three times, such that folds of stage two added to the deformation produced by folds of stage one, and folds of stage three further deformed both folds of stages one and two. It's as if the whole mass were churned in a taffy machine. The rocks throughout the Highlands Adirondacks may well have been taffy-like when they were deformed at perhaps as much as 15 miles depth, because they were subjected to such high temperatures and pressures. The folding is well-outlined farther south by long, looping bands of rock units that trend east-west. The trends along NY 28 are less well-defined, but from one side of the Adirondacks to the other appear to have the same trend as the road: a long, convex-northward curve. In fact, the routing of the road is another good example of how geology controls human development. The road is where it is because it's easier to follow geological trends than to cut across them.

Nearly all of the roadcuts between Alder Creek and Old Forge are in gneisses of several varieties, including pinkish granitic gneisses, greenish-gray hornblende gneisses, and banded gneisses with pinkish granitic veinings. Sandy, gravelly, and sometimes bouldery roadbanks are much more common than bedrock cuts; all products of sedimentation during the lake stage of Wisconsin deglaciation.

Old Forge is New York's coldest place, where winter temperatures often drop to 40-50 degrees F. below zero. The village is at the south end of Fourth Lake in the Fulton Chain of Lakes. The chain stretches northeastward from Fourth to Eighth lakes along a series of interconnected depressions, which continue on to Raquette, Utowana, Eagle, Blue Mountain, and Durant lakes. There seems little doubt that all of these lakes were interconnected during the glacial lake stage, at which time they all drained to the east into the Hudson River. Only the easternmost, Lake Durant, now drains to the Hudson, while Blue Mountain, Eagle, Utowana, and Raquette lakes drain northward to the St. Lawrence, and the Fulton Chain drains westward to the Black River, and then to Lake Ontario. Now drainage divides of glacial debris are between Blue Mountain and Durant lakes, and between Raquette and Eighth lakes.

An excellent place to survey the Fulton Chain is from the summit of Bald Mountain north of Old Forge on the west side of Fourth Lake. From there, the geological positioning of the lakes is obvious: all of the lakes from Fourth to Eighth lie at the base of scarps formed in gneisses with northwest-dipping layers. The route between Old Forge and Blue Mountain Lake (36 miles) skirts nearly all of these lakes. Aside from the summer cottages and resort developments on the lakes, it is pretty wild and densely wooded country. Again, a variety of gneisses is exposed in the many small roadcuts. Regional geological structures curve around from northeast to east and the road follows this trend. The view on this approach of guardian Blue Mountain hovering over Blue Mountain Lake is a lovely one.

View northeast from Bald Mountain along the Fulton Chain of lakes profiling the scarps (right) and the gentler dip slopes (left) of the mountains

NY 28:
Blue Mountain Lake—Warrensburg
49 mi./79 km.

Immediately east of Blue Mountain Lake, NY 28/30 passes over a low divide on glacial drift that narrowly separates the drainage basins of the St. Lawrence River and the Hudson River. Lake Durant, on the east side of the divide, drains eastward into Rock River, then to Cedar River, and finally to the Hudson, only 13 miles from Blue Mountain Lake. Glacially induced drainage changes are common in the Adirondacks, but this is probably the most striking and delicately balanced one of all.

The long roadcut near the Lake Durant parking area is in the Lake Durant formation, here principally a pinkish and greenish gneiss with strong layering and thin dark banding.

Between Lake Durant and Indian Lake village (11 miles), east of the road, note the pronounced asymmetry of Stark Hill and Ledge Mountain, a product of glacial scour and plucking. When thick ice sheets advanced over the mountains on each of four, widely separated glaciations of the Ice Age, they scoured and smoothed the upstream sides of the hills while rock of the unsupported downstream sides was broken off along fractures or formation boundaries—plucked by the sheer weight of the ice. The predominance of south-facing, jagged cliffs throughout this region of the Adirondacks leaves little doubt of the southward advance of the ice sheets.

Indian Lake lies in one of the most prominent northeast-trending fracture zones of the Adirondacks. The lake is almost perfectly aligned with Jessup Valley to the south and Lake Abanakee and the

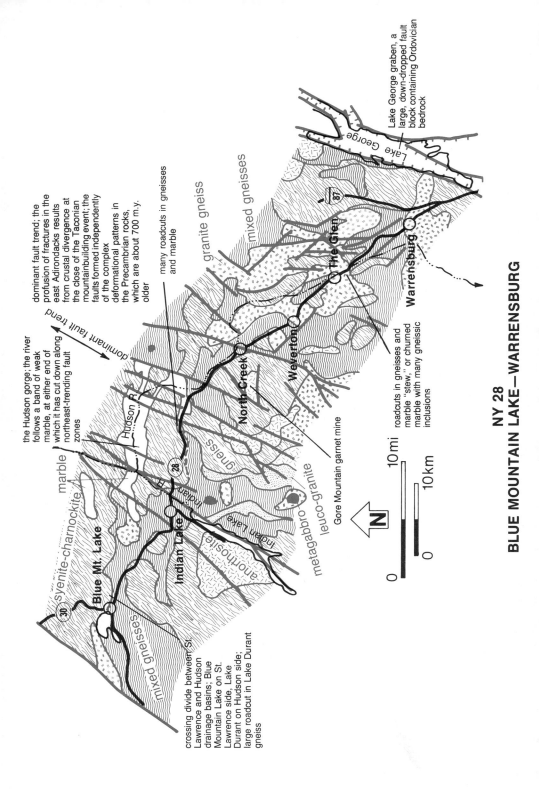

NY 28
BLUE MOUNTAIN LAKE—WARRENSBURG

the Hudson gorge; the river follows a band of weak marble, at either end of which it has cut down along northeast-trending fault zones

dominant fault trend; the profusion of fractures in the east Adirondacks results from crustal divergence at the close of the Taconian mountainbuilding event; the faults formed independently of the complex deformational patterns in the Precambrian rocks, which are about 700 m.y. older

dominant fault trend

many roadcuts in gneisses and marble

Hudson R.

marble

Blue Mt. Lake

syenite-charnockite

mixed gneisses

crossing divide between St. Lawrence and Hudson drainage basins; Blue Mountain Lake on St. Lawrence side, Lake Durant on Hudson side; large roadcut in Lake Durant gneiss

Indian Lake

Indian R.

Indian Lake

anorthosite

metagabbro

leuco-granite

Gore Mountain garnet mine

North Creek

granite gneiss

mixed gneisses

Wevertown

roadcuts in gneisses and marble "stew," or churned marble with many gneissic inclusions

The Glen

Warrensburg

Lake George

Lake George graben, a large, down-dropped fault block containing Ordovician bedrock

N

0   10 mi

0   10 km

Rafting through the Hudson Gorge
Courtesy Hudson River Rafting Co. of North Creek, New York.

Indian and Hudson rivers to the north. A ramrod-straight, four-mile section of the Cedar River follows another parallel fracture zone two miles west of Indian River.

The Hudson makes a peculiar right-angle bend four miles northeast of Indian Lake village near the Indian River junction and flows eastward out of the fault zone. It continues east for about 9 miles through the spectacular Hudson Gorge, carved in a band of marble, and then turns south for four miles to the hamlet of North River along another fault zone. The gorge is one of those hidden pockets of beauty virtually inaccessible to all but those willing to brave the turbulent river. In recent years, commercial rafting outfits have made this a possibility for large numbers of people during the high-water spring months. Most of the outfits have temporary headquarters in the North Creek-North River area. Rafts and rafters are transported to the Indian River just below the Lake Abanakee dam. That first three miles down the Indian is a wild ride of constant white water, dropping a total of 300 feet to the Hudson. The Hudson Gorge itself is marked by long stretches of tremendous rapids alternating with swift-moving flatwater. Over much of the way, banded marble is visible at water level on the south side, with layers dipping into the slope, and capped by carbonate-loving cedar trees. The rock of the north bank is mostly hard gneiss, so it is obvious that the river follows the contact between the two rock units.

The Blue Ledge is a hauntingly beautiful, 300-foot-high precipice at a sharp bend in the gorge where most of the river-runners stop for lunch. The name derives from the upper part of the cliff, which is moss-covered, and shimmers a gunmetal blue-gray as soilwater flows over it. The most fearsome rapids of all come just beyond the bend.

The route between Indian Lake village and North River (14 miles) boasts many excellent roadcuts and lovely open mountain scenery. Most of the roadcuts are along the steep incline at the North River end, and include such rocks as dark, banded gneisses, garnet-bearing hornblende gneisses like those of the Gore Mountain garnet mines, grayish gneisses, pinkish granite gneisses, and marbles. The marbles here are like "rock stew," with chaotic assortments of angular gneiss blocks suspended within them in swirling patterns. Actually, the marbles are rather intimately interlayered with gneisses, and all were subjected to the same intense metamorphism and deformation during the Grenville mountain-building event. The high temperature and pressure caused the marbles to deform like taffy, while the more brittle gneisses were broken into pieces that were swept along with the flow.

Pathfinder Camp roadcut 8 miles east of Indian Lake village on NY 28. Banded gneiss overlies marble "stew" (white) that contains numerous disoriented gneiss blocks, indicating that the marble was very plastic during metamorphism and deformation.

The road follows the Hudson River between North River and North Creek (5 miles). This is the section where most of the action takes place in the Hudson River Derby every spring, the most popular white water event in New York. At that time, every square inch of roadside is occupied by cars, tents, people, and boats.

North Creek is near the world-famous Gore Mountain garnet deposits which, at this writing, supply nearly 90% of the world's demand for industrial garnets. The garnet is ideal for abrasives because it is very hard and it breaks up in knife-sharp, angular grains, no matter how finely it is crushed. The main ore occurs as giant red garnets, sometimes two feet in diameter, set in black, shiny metagabbro—a spectacular rock. Many of the roads in the area sparkle in the sunlight because the crushed rock residue from ore processing has been used as asphalt aggregate. The glitter comes from shiny cleavage faces of the black mineral, hornblende. Besides Gore Mountain, other garnet deposits have been mined near North Creek and in the Garnet Lake and Thirteenth Lake regions. Mineral collectors favor these deposits because, in addition to garnet, a wide variety of other minerals occur in them, sometimes in museum-quality crystals.

The road follows the Hudson only part of the way between North Creek and Warrensburg (18 miles). A few roadcuts in this stretch display dark metagabbro, banded gneisses, calc-silicate rocks made up primarily of calcium silicate minerals, marble "stew," and gray and pinkish gneisses.

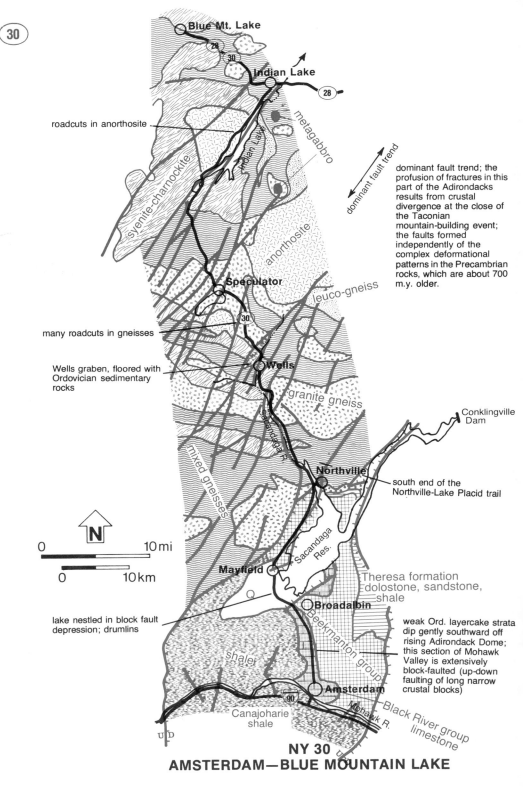

Blue Mt. Lake

28
30

Indian Lake

28

roadcuts in anorthosite

metagabbro

Indian Lake

syenite-charnockite

anorthosite

dominant fault trend

dominant fault trend; the profusion of fractures in this part of the Adirondacks results from crustal divergence at the close of the Taconian mountain-building event; the faults formed independently of the complex deformational patterns in the Precambrian rocks, which are about 700 m.y. older.

Speculator

leuco-gneiss

30

many roadcuts in gneisses

Wells graben, floored with Ordovician sedimentary rocks

Wells

granite gneiss

Conklingville Dam

Sacandaga R.

mixed gneisses

Northville

south end of the Northville-Lake Placid trail

N

0   10mi

0   10km

Sacandaga Res.

Mayfield

Theresa formation dolostone, sandstone, shale

lake nestled in block fault depression; drumlins

Broadalbin

Beekmantom group

weak Ord. layercake strata dip gently southward off rising Adirondack Dome; this section of Mohawk Valley is extensively block-faulted (up-down faulting of long narrow crustal blocks)

shale

90

Amsterdam

Canajoharie shale

U D

Mohawk R.

Black River group limestone

NY 30
AMSTERDAM—BLUE MOUNTAIN LAKE

NY 30:
Amsterdam—Blue Mountain Lake
93 mi. / 150 km.

The city of Amsterdam is on the steep north bank of the Mohawk Valley, a few miles south of the southern exposed boundary of the Adirondack dome. The river occupies center stage in the wider Mohawk Lowlands that separate the Adirondacks from the Allegheny Plateau. Amsterdam is near the apex of a broad triangle at the eastern end of the lowlands where they blend into the Hudson Lowlands. All are floored by weak Ordovician strata, and here they dip, or slope, gently southward off the Adirondack dome (see NY 37: Malone—Ogdensburg discussion). Deep erosion of these rocks produced the Allegheny scarp in the resistant Helderberg/Onondaga limestones, not only in this region, but also on the east side of the Helderberg Mountains facing the Hudson Valley.

Much of the work of excavating the Mohawk Valley was accomplished, not by the present river, but by the much larger Iro-Mohawk during Wisconsin deglaciation (see Ice Age chapter). The valley then served as the principal escape route for virtually all of the glacial meltwaters impounded in all of the juvenile Great Lakes. At an earlier time, ice that lapped up onto the Allegheny Plateau blocked the valley while the Susquehanna River system channeled the meltwaters. At a later time, the St. Lawrence River became the exclusive escape route when it was free of ice, and the Mohawk eventually diminished to its present status.

Block or up-down faulting is characteristic of the Mohawk Valley, as it is of the southern and eastern Adirondacks and Champlain Valley (see Interstate 90: Syracuse—Massachusetts border). Most of the faults here trend northeasterly and extend from deep within the Precambrian terrane at least to the south side of the river; a few have been traced to the Allegheny scarp, where they cut the Ordovician rocks but not the overlying Silurian or Devonian strata. It is fairly certain, therefore, that the faults are of late Ordovician age, formed during a crustal divergence in the waning stages of the Taconian mountain-building event (see Plate Tectonics section). Among the faults are a number of elongate, downdropped blocks referred to by the German term "graben," meaning grave. Ordovician and Cambrian rocks are almost invariably well-preserved in the graben, while the upthrown blocks on either side are worn down to the Precambrian basement rocks. The graben are natural lake sites, and one of the largest is the Sacandaga graben directly north of Amsterdam, containing Great Sacandaga Lake.

NY 30 rises about 450 feet in the first two miles north of the Mohawk River, passing through Amsterdam. The way is steep, and views south across the valley reveal its great size as compared to the modern river that flows through it.

Between the top of the Amsterdam slope and Broadalbin (10 miles), there is a very gradual rise in elevation over middle Ordovician Trenton/Black River limestones. At Broadalbin, you cross a low fault-line scarp marking the southeastern boundary of the Sacandaga graben. Between Broadalbin and Mayfield (6 miles), you cross to the opposite side of the graben at the south end of the lake on a flat sediment plain dotted with drumlins, several of which are more than 200 feet high. Between Mayfield and Sacandaga/Northville (10 miles), the road follows the scenic shore of the lake, with the 1000-foot high scarp of the Noses fault rising rather abruptly to the west (see Interstate 90: Syracuse—Massachusetts border discussion). You cross the fault just south of the Northville bridge, and the crossing is marked by the transition from roadcuts in grayish, well-bedded dolostones of the Cambrian Theresa formation or Beekmantown group in the graben, to roadcuts in banded gneisses on the upthrown side of the fault. The upthrown side also displays the rugged, knobby topography typical of the Highlands Adirondacks region, a reflection of the complex geology and variety of metamorphic rock types.

Among Adirondack lakes, Great Sacandaga has an unusual history. The modern lake is really a man-made reservoir behind the Conklingville dam at the north end of its long, northeastern arm, just three miles from the Hudson River. But it is an almost exact duplicate of a postglacial lake that had since drained itself. Before the Ice Age, the ancestral Sacandaga River drained most of the southern Adirondacks, passing through the lake basin to the ancestral Mohawk, which, in turn, emptied into the ancestral Hudson. During Wisconsin deglaciation, the south end of the basin between Broadalbin and Gloversville was clogged with glacial sediments and the lake filled up behind them. The water level rose ever higher, until it spilled over the divide at Conklingville into the Hudson. Eventually the divide was worn away and the lake drained completely. The Conklingville dam was completed in 1930 to stem the devastating floods that plagued the Hudson River valley.

Between Northville and Wells (17 miles), the road goes upstream along the Sacandaga River, whose course follows several fault zones. En route, you pass roadcuts in gneiss, with steeply dipping layers. North of Hope, you also pass through a small graben floored with Beekmantown dolostones, but exposures are hard to find. Wells lies in the Wells graben, containing Beekmantown dolostones, Trenton/ Black River limestones and Canajoharie black shales; there are exposures of the latter across Algonquin Lake from the village.

Grabens like these, deep within the Precambrian terrane of the Adirondacks, reveal a great deal about the geologic history of the region, as follows. By early Paleozoic time the Grenville-age (1100 million years) ancestral Adirondacks were worn down to a surface of such low relief that it was eventually submerged beneath the waters of a shallow inland sea. It was in this sea that the original Cambrian and Ordovician sediments were laid down. By the end of middle Ordovician time, most of the Adirondack region was blanketed with sedimentary strata. The block-faulting occurred during the late Ordovician Taconian mountain-building event, setting the stage for long-term preservation of the sedimentary rocks in the graben. Stripping of these same strata from the high places was completed much later, in fact only during the last few million years when the Adirondack dome began to rise.

The road trends diagonally across numerous other northeast-trending fault zones between Wells and Indian Lake village (38 miles). Indian Lake is hollowed out of two of these zones. There are many roadcuts in a variety of gneisses, some of which show much folding and complex deformation. All of the bedrock, in fact, from here northward nearly to Malone is Precambrian; there are no more Paleozoic-floored graben along NY 30. The Snowy Mountain dome, a large mass of Adirondack anorthosite, intersects the western shoreline over two-thirds of the length of the lake (see Whiteface Mountain Memorial Highway discussion). Marbles are rare in most of the Highlands Adirondacks. They are abundant, however, interlayered with gneisses, in a small region northeast of Indian Lake. The course of the Hudson River there, in fact, appears to be controlled by easily-eroded marble bands.

The drive between Indian and Blue Mountain lakes (13 miles) is one of the prettiest in the Adirondacks. Owing to road widening and resurfacing in the 1970s, there are now many fresh, landscaped roadcuts, and views of the mountains are spectacular. The Blue Mountain Lake setting is idyllic, with the monolithic dome of Blue Mountain looming large over its east side. Perhaps better than any other peak, this one illustrates the domical morphology so characteristic of Adirondack mountains. It results from the streamlining action of overriding ice, coupled with the massive character of the bedrock and sparse fracturing.

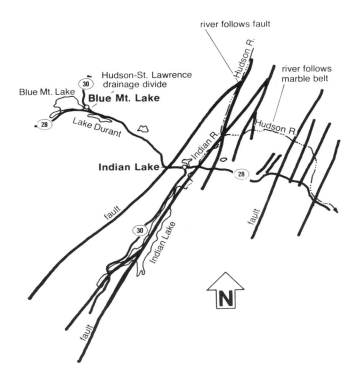

Fault control of the position of Indian Lake and the Hudson River. The E-W section of the Hudson follows a weak marble belt.

NY 30:
Blue Mountain Lake—Malone
95 mi. / 154 km.

Blue Mountain Lake, with its guardian Blue Mountain hovering nearby, is surely a gem among the Adirondacks' 2000-odd lakes and ponds; and its unusual beauty has contributed to its development as a remote cultural center. Here, for example, are the magnificent Adirondack Museum, devoted principally to Adirondack history and culture, and the Adirondack Lakes Center for the Arts, which sponsors plays and concerts, arts and crafts, courses in photography, printing, and sculpture, and features artists in residence during the summer. Happily so far, there seems to be very little effort to enlarge on these activities or to over-commercialize their idyllic setting.

Blue Mountain Lake is at the headwaters of the Raquette River. Its outlet is at the west end of the lake, to Eagle, then Utowana, then Raquette Lakes. The drainage from Raquette Lake then goes in the exact opposite direction to the east, first to Forked Lake, then to Long Lake, and then continues generally northward to the St. Lawrence River. This is a peculiar reversal of course that was glacially influenced. The preglacial drainage route was to the Hudson River, 13 miles directly east, a route now blocked by glacial drift at the east end of the lake. Lake Durant, only a stone's throw beyond the barrier, drains to the Hudson; the barrier, therefore, is distinguished as the drainage divide between two great river systems, the St. Lawrence and the Hudson.

343

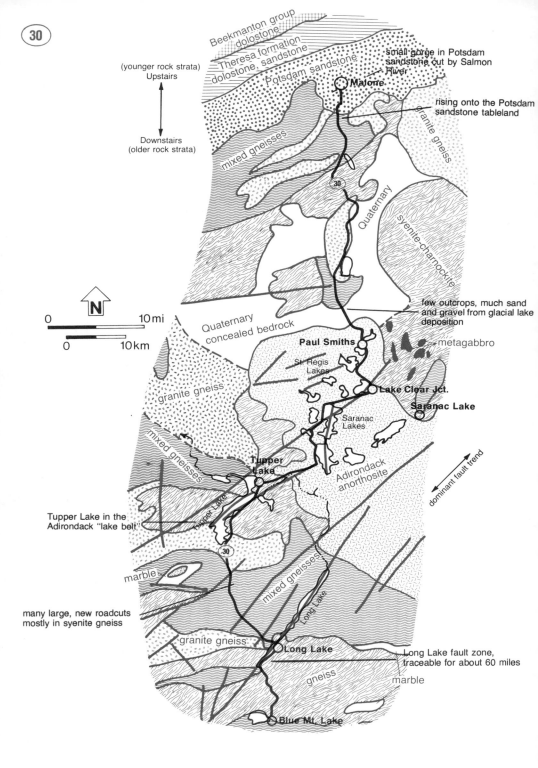

(younger rock strata)
Upstairs

Downstairs
(older rock strata)

Beekmanton group
dolostone
Theresa formation
dolostone, sandstone
Potsdam sandstone

small gorge in Potsdam
sandstone cut by Salmon
River

Malone

rising onto the Potsdam
sandstone tableland

granite gneiss

mixed gneisses

Quaternary

syenite-charnockite

few outcrops, much sand
and gravel from glacial lake
deposition

Quaternary
concealed bedrock

metagabbro

Paul Smiths

St Regis
Lakes

Lake Clear Jct.

Saranac Lake

granite gneiss

Saranac
Lakes

mixed gneisses

dominant fault trend

Tupper Lake

Adirondack
anorthosite

Tupper Lake in the
Adirondack "lake belt"

marble

many large, new roadcuts
mostly in syenite gneiss

mixed gneisses

granite gneiss

Long Lake

Long Lake fault zone,
traceable for about 60 miles

gneiss

marble

Blue Mt. Lake

0   10 mi
0   10 km

N

NY 30
BLUE MOUNTAIN LAKE—MALONE

The drive on NY 30 between Blue Mountain Lake and Long Lake (11 miles) now has many fresh roadcuts because of road widening in 1983. At its northern end, the route follows the scenic east shore of Long Lake for three miles.

Long Lake has been called the "javelin of the mountains," because it is so long (12 miles), straight, and narrow (average half-mile); it appears more like a river than a lake. The fault zone that was scooped out by streams and glaciers to form the lake basin has been traced for about 40 miles beyond the southwest end of the lake, and 10 miles beyond the northeast end. The northern part of the lake, visible where the road crosses it, is strikingly beautiful, studded with wooded islands and seeming to go on forever.

The pretty, northern end of the route between Long Lake and Tupper Lake (22 miles) has also been recently widened and resurfaced, making huge roadcuts. Most are in dark green syenite gneiss, a rock closely associated with the Adirondack anorthosite massif, whose western boundary lies a few miles to the east.

Tupper Lake is at the southwestern end of the Saranac intramontane basin, an anomalous lowland of the mountains blessed with so many canoeable lakes that it is often referred to as the "lake belt." (see NY 86: Lake Clear Junction—Lake Placid—Jay). There are about 50 lakes of a mile or more diameter, including the Tupper Lakes, the Saranac Lakes, and the St. Regis Lakes, and well over a hundred smaller ones. The basin trends northeasterly and is about 35 miles long by 10 to 15 miles wide. The depression is probably part of an unusually large, composite, graben structure.

The road follows the lake basin between Tupper Lake and Paul Smiths (27 miles) and weaves its way among the lake clusters. Relief throughout is limited to about 460 feet, and there are many sandy flats. As you may have noticed, mixed sand and gravel deposits are abundant in all the Adirondack valleys. They are mostly products of the late stage of Wisconsin deglaciation, when the valleys were filled with meltwater and received great quantities of sediments from inflowing streams.

Between Paul Smiths and Malone (37 miles), you pass over rolling country with abundant sand/gravel roadbanks, a few roadcuts that display a variety of gneisses, and numerous open views of densely wooded foothills. There are also a lot of boggy wetlands, as there are throughout the Adirondack region. Peat bogs are common here because glaciation left many poorly-drained depressions. Some are kettle holes where stagnant blocks of ice were buried in outwash sediments during deglaciation; most are simply sites of former meltwater lakes or ponds in which were deposited impermeable layers of glacial "flour," called "varves" All are characterized by poorly integrated inflowing and outflowing drainage. Under these conditions, wetlands vegetation takes over, encroaching upon open water and slowly filling the depressions with peat, the partly carbonized residium of roots, tree trunks, twigs, seeds, shrubs, stems, mosses, and other vegetable matter. An oxygen-poor aquatic environment is essential to peat formation; otherwise plant material would rot.

There are Adirondack bogs in every stage of development, some with a lot of open water, some completely filled in and virtually dry, some partially filled with very little remaining open water and containing a floating vegetable mat that extends inward from the shores. The latter reveal much about their geologic and biologic evolution. There you will often see swamp trees, like black spruce, tamarack and cedar around the outer periphery growing in the peat rather than on solid ground, then inward toward open water, heath shrubs like alder, and then a wet spongy mat of sedges, grasses, and moss with a lot of small marsh plant growth, often including insectivores. Walking on the mat is like walking on a water bed.

In the last 4 miles to Malone, the road levels off on a tableland of Potsdam sandstone, which dips, or slopes, gently northward off the Adirondack dome. Excellent outcrops of the Potsdam are visible in the small gorge of the Salmon River in the center of Malone (see NY 37: Malone—Ogdensburg discussion).

NY 3:
Watertown—Tupper Lake
105 mi./170 km.

MOUNTING THE ADIRONDACK DOME

Between Watertown and Carthage (22 miles), this route follows the Black River upstream, skirting the Tug Hill, with its northern scarp rising immediately south of the river. The river here, as on the east side of the Tug Hill, is hard against the scarp, with open valley on the opposite side (see NY 12: Watertown—Boonville discussion). Grayish Black River limestone is exposed in the river banks where the road crosses over in two places. The levelness of the road in this stretch is a reflection of the underlying, nearly flat-lying strata. Actually, the beds dip to the southwest off of the Frontenac arch, the northwest-trending ridge of Precambrian rocks exposed in the Thousand Islands; but the dip angle is almost imperceptible.

Between Carthage and Natural Bridge (9 miles), the road turns northeastward and enters the Precambrian terrane, passing several roadcuts in a variety of gneisses. The route lies very close to the faulted boundary between the Highlands Adirondacks to the southeast and Lowlands Adirondacks to the northwest. The boundary snakes back and forth for 70 miles in a northeasterly direction between Carthage and Parishville, and dives under alluvium and Paleozoic strata at either end.

Natural Bridge is so-named because it is the site of a small cave in the Gouverneur marble on the Lowlands side of the Highlands-

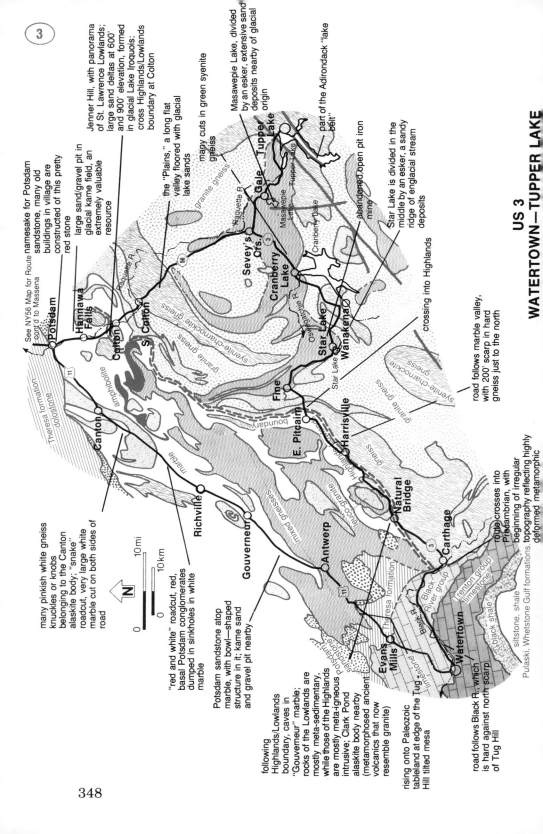

US 3
WATERTOWN—TUPPER LAKE

See NY56 Map for Route cont'd to Massena

namesake for Potsdam sandstone, many old buildings in village are constructed of this pretty red stone

large sand/gravel pit in glacial kame field, an extremely valuable resource

Jenner Hill, with panorama of St. Lawrence Lowlands; large sand deltas at 600' and 900' elevation, formed in glacial Lake Iroquois; cross Highlands/Lowlands boundary at Colton

Masawepie Lake, divided by an esker, extensive sand deposits nearby of glacial origin

part of the Adirondack "lake belt"

the "Plains," a long flat valley floored with glacial lake sands

many cuts in green syenite gneiss

abandoned open pit iron mine

Star Lake is divided in the middle by an esker, a sandy ridge of englacial stream deposits

road follows marble valley, with 200' scarp in hard gneiss just to the north

crossing into Highlands

many pinkish white gneiss knuckles or knobs belonging to the Canton alaskite body; "snake" roadcut, very large white marble cut on both sides of road

"red and white" roadcut, red, basal Potsdam conglomerates dumped in sinkholes in white marble

Potsdam sandstone atop marble, with bowl-shaped structure in it; kame sand and gravel pit nearby

following Highlands/Lowlands boundary, caves in "Gouverneur" marble; rocks of the Lowlands are mostly meta-sedimentary, while those of the Highlands are mostly meta-igneous intrusive; Clark Pond alaskite body nearby (metamorphosed ancient volcanics that now resemble granite)

rising onto Paleozoic tableland at edge of the Tug Hill tilted mesa

road follows Black R. which is hard against north scarp of Tug Hill

route crosses into Precambrian, with beginning of irregular topography reflecting highly deformed metamorphic

Pulaski, Whetstone Gulf formations: siltstone, shale

black shale

Trenton group limestone

Black River group

Theresa formation

Theresa formation dolostone

granite gneiss

syenite-charnockite gneiss

granite gneiss

marble

amphibolite

mixed gneisses

Highlands gneiss

leuco-granite

Potsdam sandstone

limestone

syenite-charnockite gneiss

Boundary

Potsdam, Hannawa Falls, Colton, S. Canton, Canton, Richville, Gouverneur, Antwerp, Evans Mills, Watertown, Carthage, Natural Bridge, Harrisville, E. Pitcairn, Fine, Star Lake, Wanakena, Cranberry Lake, Sevey's Crs., Gale, Tupper Lake

Raquette R., Oswegatchie R., Cranberry Lake, Masawepie Lake, Tupper Lake, Star Lake, Black R.

10 mi / 10 km N

348

Lowlands boundary. The cave is the underground channel for some of the Indian River discharge and must be seen by boat. Small caves like this are a common feature of the Lowlands Adirondacks, especially along the rivers, which prefer to follow the easily eroded marbles rather than the harder silicate rocks.

The route continues to straddle the Highlands-Lowlands boundary between Natural Bridge and Harrisville (11 miles), narrowly skirting the east side of the Lake Bonaparte "alaskite" body en route. Alaskite is defined as granite that is relatively free of dark mineral constituents, such as pyroxene, hornblende, or biotite. Large, oval-shaped masses like the Lake Bonaparte body are dispersed throughout the Lowlands among the marbles and gneisses. All such bodies were originally considered to be igneous, or molten rock, intrusions that forced their way between the folded metamorphic rock layers and solidified. This interpretation is no longer popular among geologists. Instead, the alaskites are now being lumped together as a sheetlike unit that was interstratified, folded, and metamorphosed along with the marbles, metasedimentary gneisses, and other, relatively minor rock types (see NY 37, 12, 12E: Ogdensburg—Cape Vincent—Watertown discussion). The original rock is thought to have been volcanic, mainly debris from explosive eruptions like that of the 1980 eruption of Mt. St. Helens in Washington State. These materials accumulated in horizontal layers that covered the sedimentary strata over a wide area and were later covered by new sediments. The original volcanic mineralogy and texture are now almost completely changed by the intense Grenville metamorphism.

The author believes that the granitic rocks in the vicinity of Alexandria Bay, for which this "unit" has been named, are incorrectly lumped together with the alaskites. Instead, they appear to be true granitic intrusions that invaded the country rock as magma and never broke through to the surface.

The dominant outcrop and topographic trend of the Lowlands Adirondacks is northeasterly, reflecting a pronounced large-scale wrinkling of strata in that direction. The deformation, however, is far more complex; geologists now believe there are at least three, and maybe four, sets of folds of different orientation and intensity superimposed on each other that account for the wild assortment of outcrop loops, "fishhooks," and ovals. The same deformational patterns may also characterize the Highlands, but they are better defined in the Lowlands because of the more stratified nature of the original rocks. By contrast, many of the Highlands rocks appear to have begun as large, irregular igneous intrusive bodies.

Most of the roadcuts between Natural Bridge and Harrisville are dark greenish-gray, rusty-weathering, syenite gneiss, a rock of quartz-poor granitic composition that forms a large, elongate body adjacent to the Highlands-Lowlands boundary.

You cross the west branch of the Oswegatchie River at Harrisville. From here, the river works its devious way northward to enter the St. Lawrence River at Ogdensburg. In crossing the Lowlands, it is forced repeatedly to follow marble valleys northeastward or southwestward between hard gneiss ridges, so it must travel many times the straight line distance to its destination (see US 11: Potsdam—Watertown discussion).

The route continues near the Highlands—Lowlands boundary between Harrisville and East Pitcairn (7 miles). Just north of Harrisville, it enters a marble valley, with an impressive 200-foot high erosional scarp on gneiss north of the road. You cross into the Highlands as the road veers to the east between East Pitcairn and Fine (5 miles). At Fine, the road intercepts the east branch of the Oswegatchie, which has its main headwaters at Cranberry Lake.

Star Lake, 8 miles southeast of Fine, is a uniquely beautiful small lake that is split right down the middle by an esker, one of those long, winding ridges of sand of glacial origin. There are numerous Adirondack lakes with eskers in their midst, but no others so evenly divided as this one. All of the eskers formed in the same way, as sandy, gravelly deposits along streams that flowed on, or even through tunnels in, the wasting Wisconsin ice sheet. When the ice finally melted, the deposits were laid more or less gently on the ground directly below. Their sinuous courses make them appear, from the air, like rivers of sand.

Rainbow Lake esker from Meenahga Mountain, forms a long narrow island covered with sand-loving white pines.
Courtesy Kendall/Hunt Publishing Co.

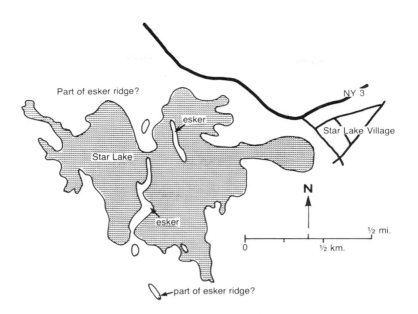

Map of Star Lake, a probable kettle lake with a dividing esker ridge of sand and gravel down the middle

Just east of Star Lake is the enormous open pit iron mine formerly operated by the J and L Steel Corporation. The primary ores mined here are magnetite and martite. Similar deposits were extensively mined in numerous parts of the Adirondacks in the past. Magnetite has the unique property of being magnetic, which greatly facilitates its separation from the crushed ore rock simply by using powerful electromagnets. Martite is really magnetite that has been replaced by another iron oxide mineral, hematite, while it retains the original form of the magnetite. The mine is an excellent mineral collecting locality where at least 21 different minerals have been found. The large area of barren, dusty ground around the mine is made up of tailings, the residue of crushed rock that must be discarded after the ore is extracted from it.

Cranberry Lake, 14 miles east of Star Lake, is a much larger lake that is even more star-like in outline, with long arms radiating in several directions. Its unusual shape results from flooding of side valleys after construction of a dam on the east branch of the Oswegatchie near Cranberry Lake village. The upper reach of the east branch of the Oswegatchie now flows into the Wanakena arm of the lake.

There are many new, large roadcuts between Cranberry Lake and

Seveys Corners (8 miles) that display a variety of gneisses, including dark greenish syenites, pinkish granites, and banded gneisses, the gneissic layering in one cut is transected by extremely coarse-grained, pinkish pegmatite, an igneous intrusive rock that often contains exotic minerals. There is also an abundance of sand in the roadbanks, as there is in all Adirondack valleys. The sand was deposited during the lake stage of Wisconsin deglaciation when so many valleys were filled with meltwater lakes and ponds. Many of the boggy wetlands that are so characteristic of the region are the partially filled meltwater basins.

The route between Seveys Corners and Tupper Lake (18 miles) is more of the same, but the bedrock roadcuts are not as fresh. The route approximately follows the course of the Raquette River, one of the most dammed-up small rivers in the eastern United States (see NY 56: Massena—Seveys Corners discussion). En route, at Gale, you pass the Massawepie Boy Scout Camp where another fine esker stretches for miles and forms a sandy, sharp-crested, barrier between several lakes.

Visibility opens up on the long descent to Tupper Lake from the west. The lake is one of many in the so-called Saranac intramontane basin, a large, anomalous lowland of the mountains that stretches for about 35 miles in a northeast direction and is 10 to 15 miles wide (see NY 30: Blue Mountain Lake—Malone discussion).

NY 3:
Tupper Lake—Plattsburgh
70 mi./113 km.

DOWN THE EAST SIDE OF THE ADIRONDACK DOME

This route trends northeastward across the Highlands Adirondacks and then descends to the Champlain Lowlands at Plattsburgh. Between Tupper Lake and Saranac Lake (22 miles), the road cuts across the northwestern salient of the Marcy massif of the Adirondack anorthosite, which underlies all of the High Peaks region to the east. Several old roadcuts expose the anorthosite en route. At either end of this section of highway is dark greenish gray syenite gneiss at the borders of the anorthosite body. Mt. Morris, the site of Big Tupper ski area directly south of Tupper Lake village, is carved from this same rock. The anorthosite itself is believed to have originated as igneous melt that welled up from great depth along two or three pipes and then spread out, forcing its way among the country rocks to form an enormous mushroom-shaped mass. The igneous character of the mass is now partly obscured by overprinting of metamorphism, including much crushing and recrystallization that accompanied the Grenville mountain-building event, 1100 million years ago (see also Whiteface Mountain Memorial Highway discussion). This possibly occurred at depths of up to 15 miles, under extreme temperatures and pressures, in the deep core of the ancestral Adirondacks. Other anorthosite bodies of similar origin are in the Grenville Province of Canada, most notably in Labrador. Present exposure of these bodies at the surface means, of course, that erosion has removed several miles of rock.

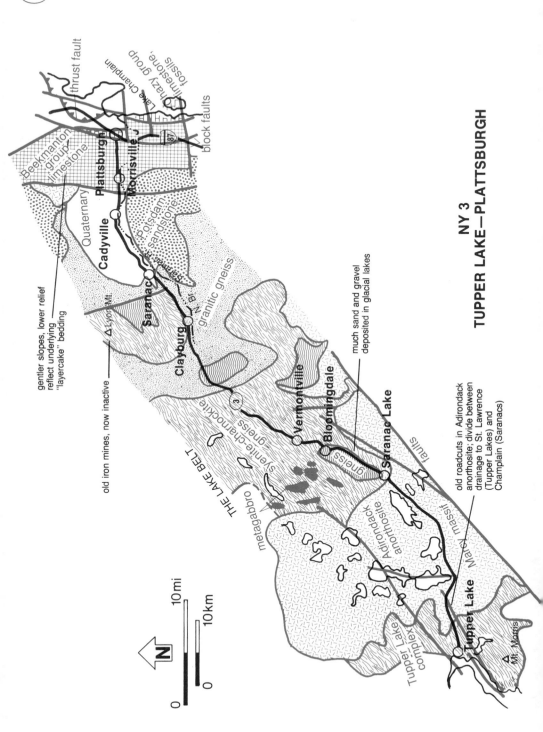

NY 3
TUPPER LAKE—PLATTSBURGH

thrust fault

Lake Champlain

Chazy group

limestone, fossils

block faults

Beekmantown group limestone

Quaternary

Potsdam sandstone

Plattsburgh

Morrisville

87

Cadyville

Saranac River

△ Lyon Mt.

Saranac

N. Br.

Clayburg

granitic gneiss

3

Vernonville

Bloomingdale

Saranac Lake

gneiss

syenite-charnockite gneiss

metagabbro

THE LAKE BELT

Adirondack anorthosite

Tupper Lake complex

Tupper Lake

△ Mt. Morris

Marcy massif

faults

gentler slopes, lower relief reflect underlying "layercake" bedding

old iron mines, now inactive

much sand and gravel deposited in glacial lakes

old roadcuts in Adirondack anorthosite; divide between drainage to St. Lawrence (Tupper Lakes) and Champlain (Saranacs)

N

0   10 mi

0   10 km

The many lakes in this section of the Adirondacks have earned it the name of "lake belt." Tupper Lake is artificially enlarged by the Sitting Hole dam, the highest in the 19 Raquette River dams under control of Niagara-Mohawk Power Corporation, but most of the other lakes in the belt are in their natural state. The larger ones include the numerous Saranacs and St. Regis lakes. Altogether there are about 50 lakes a mile or more in diameter and well over one hundred smaller ones. Nearly all are canoeable, so this is really a canoeist's paradise. There is a drainage divide between Tupper Lake and the Saranacs; Tupper Lake drains via the Raquette River to the St. Lawrence while the Saranac and St. Regis lakes drain northeastward to Lake Champlain via the Saranac River.

The route between Tupper Lake and Saranac Lake is very woodsy and visibility is poor except at the Tupper Lake end. Near the other end, you cross an arm of Lower Saranac Lake.

Between Saranac Lake and Bloomingdale (7 miles) you follow the Saranac River downstream through a very sandy valley. The abundant sand here and in most Adirondack valleys was deposited in the many meltwater lakes that filled them during and after Wisconsin deglaciation. The thick, clean sand deposits may be mostly valley-filling deltas formed at the mouths of inflowing streams.

Between Bloomingdale and Riverview (17 miles), the road crosses the divide to the north branch of the Saranac River, then follows it downstream to the east for 3 miles where it joins the main branch at Clayburg. You follow the main branch most of the rest of the way to Plattsburgh.

The high mountain about 5 miles north of Clayburg is Lyon Mountain, site of the now-inactive Lyon Mountain iron mines. The ores there are classed as nontitaniferous magnetite, a type that occurs widely in Adirondack gneisses of syenitic and granitic composition, and that has been extensively mined in the past from about 120 different localities. Belonging in the same group are deposits at Benson mines near Star Lake (see NY 3: Watertown—Tupper Lake discussion), Mineville, Port Henry, Crown Point, Dannemora, Clintonville, and even Clayburg itself. Note that most of the roadcuts in the Clayburg vicinity are pinkish, fine-grained granitic gneisses. The magnetite is unique in being strongly attracted to a magnet, which makes it easy to separate from the crushed rock. Processing is also unhindered by high titanium concentration. By contrast, magnetite ores found in the Adirondack anorthosite, for example at Tahawus, contain a lot of ilmenite, a titanium-iron oxide mineral that resembles magnetite and is found intimately intergrown with it. Those deposits are not as suitable for extraction of iron and instead are mined for their titanium oxide which is used as a paint pigment. The Tahawus mines have been one of the major world suppliers of titanium oxide. It is interesting to note that some of the moon rocks returned by the Apollo astronauts are anorthosite with very high ilmenite concentration.

The foothills gradually flatten eastward from Clayburg. The Precambrian/Potsdam sandstone contact is at Saranac (9 miles from Clayburg). The nearly level landscape with gentle slopes from there eastward to Plattsburgh (15 miles) reflects the layercake stratification of the Paleozoic sedimentary rocks of the Champlain Lowlands. The rocks are actually block faulted so that their beds dip this way and that, but not steeply.

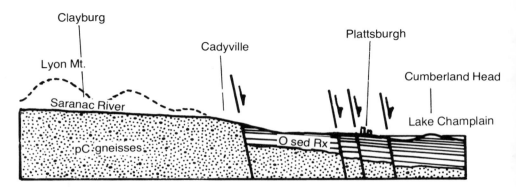

Schematic cross-section along NY 3 between Clayburg and Plattsburgh, showing block faulting.

356

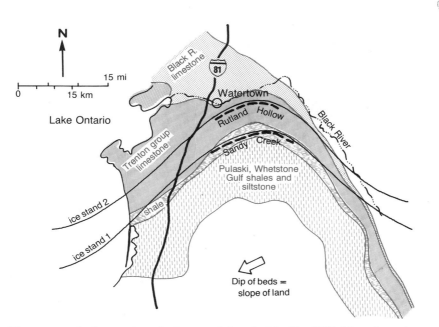

Wrap-around of retreating glacier at north end of the Tug Hill. Map shows just two hypothetical stands of Wisconsin ice marked by marginal meltwater channels at Sandy Creek (earliest) and Rutland Hollow of Mill Creek (latest).

NY 12:
Watertown—Alder Creek
57 mi./92 km.

This route skirts the northern and eastern scarp slopes of the Tug Hill plateau, very close to the Precambrian Adirondacks boundary. The Tug Hill is more accurately described as a cuesta, or tilted mesa, made up of a layercake sequence of early Paleozoic sedimentary rock strata that is tilted gently westward as a result of the domical uplift of the Adirondacks. It is this tilting that led to the erosional sculpturing of an east-facing scarp and a gentle dip slope on the opposite side, with the resistant, middle-late Ordovician Whetstone Gulf and Oswego formations serving as caprock. The northern end is truncated against the Frontenac arch, the northwest-trending Precambrian ridge exposed in the Thousand Islands, that apparently was uplifted contemporaneously with the Adirondack dome. In reality, the scarps are composites of ledges developed on several resistant beds of the middle Ordovician Black River and Trenton groups, and the Whetstone Gulf formation.

357

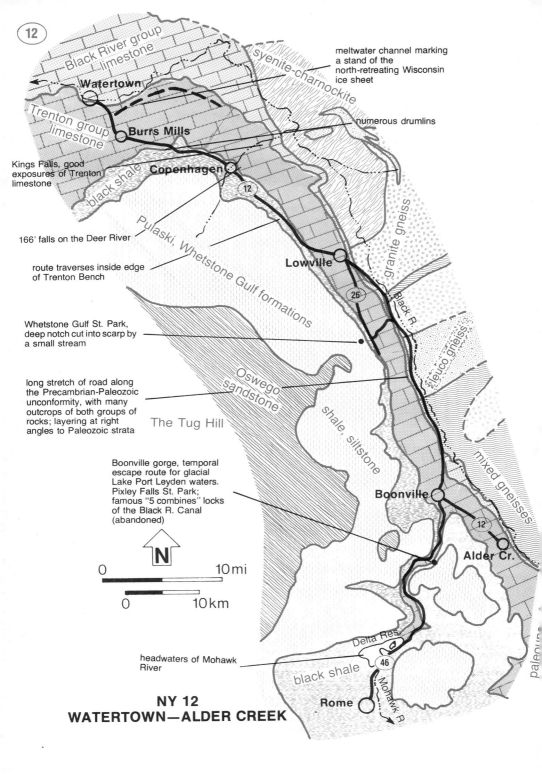

meltwater channel marking a stand of the north-retreating Wisconsin ice sheet

numerous drumlins

Black River group limestone

syenite-charnockite

Watertown

Trenton group limestone

Burrs Mills

Kings Falls, good exposures of Trenton limestone

Copenhagen

black shale

12

166' falls on the Deer River

route traverses inside edge of Trenton Bench

Whetstone Gulf St. Park, deep notch cut into scarp by a small stream

long stretch of road along the Precambrian-Paleozoic unconformity, with many outcrops of both groups of rocks; layering at right angles to Paleozoic strata

Boonville gorge, temporal escape route for glacial Lake Port Leyden waters. Pixley Falls St. Park; famous "5 combines" locks of the Black R. Canal (abandoned)

headwaters of Mohawk River

granite gneiss

Lowville

26

Black R.

leuco gneiss

Pulaski, Whetstone Gulf formations

Oswego sandstone

The Tug Hill

shale, siltstone

mixed gneisses

Boonville

12

Alder Cr.

N

0   10 mi

0   10 km

Delta Res.

46

black shale

paleozoic

Rome

Mohawk R.

NY 12
WATERTOWN—ALDER CREEK

12

The northern scarp of the Tug Hill between Watertown and Burrs Mills (4 miles) is deeply creased by meltwater channels separating erosional outliers of the plateau. Watertown's Thompson Park, just southeast of the city, is on top of one such remnant hill, capped by Trenton limestone. The channels have arcuate trends concentric with the northern rim of the plateau. They were probably initiated along the ice margin during temporary stands in Wisconsin deglaciation, while ice lobes still thrust far southward into the Black River valley to the east and Ontario basin to the west. The ice front thus wrapped around and overlapped the blunt northern end of the plateau, while the torrents of meltwater it released cut deeply into the bedrock at its base.

Between Burrs Mills and Copenhagen (9 miles), you cross the plateau in an up-dip direction, and the elevation increases gradually from 800 to 1200 feet. Numerous well-formed drumlins, a few exceeding 100 feet in height, top the surface.

Copenhagen is near the rim of one of the Trenton limestone benches mentioned earlier, where the Deer River drops 166 feet over the edge of the bench one-half mile downstream from the village. An additional 59 feet of Trenton beds is in the back wall above the falls,

166-foot falls over Trenton limestone in Deer River near Copenhagen

King's Falls on the Deer River, another step at the edge of the Tug Hill scarp, downstream from Copenhagen Falls
Courtesy Kendall/ Hunt Publishing Co.

so that the total Trenton section exposed here is 225 feet. Kings Falls, three miles farther downstream, drop over another lower step. All such falls were initiated farther downstream near the close of the Ice Age, but have since migrated upstream.

The road between Copenhagen and Lowville (13 miles) traces the inner edge of a Trenton limestone bench, along its contact with the overlying Utica shale. The roadbed is quite level over most of the way, at an elevation of almost 1200 feet. The slope immediately west, rising steeply 500-600 feet above the bench, is the main Tug Hill scarp in the Whetstone Gulf formation. Several small streams along the way have cut V-shaped notches in the scarp slope. An especially deep one a few miles south of Lowville is now the site of Whetstone Gulf State Park.

Over most of the way between Lowville and Boonville (24 miles), the road follows the base of the lowest step in the scarp slope close to the Precambrian contact. Exposures along the way include both the Precambrian basement rocks and Black River limestone that lies directly on top of them. It must be remembered that the contact surface between these two bodies of rock is an erosional "unconformity" of titanic dimensions, representing about 650 million years of missing rock record.

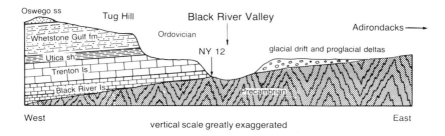

West Tug Hill · Oswego ss · Whetstone Gulf fm · Utica sh · Trenton ls · Black River ls · Ordovician · Black River Valley · NY 12 · Precambrian · glacial drift and proglacial deltas · Adirondacks → · East · vertical scale greatly exaggerated

West-east cross-section across the Black River valley, showing Trenton limestone shelf   Courtesy Kendall/Hunt Publishing Co.

Thus, the Lowville-Boonville traverse is unique in two ways: first, because it lies along a temporal boundary of tremendous significance; secondly, because it is a dynamic boundary that marks the current limit of the ongoing erosional stripping of the Adirondack dome. With these facts in mind as you travel the route, note the "gneiss knuckles" jutting from the soil east of the road, and Black River limestone outcrops west of the road. Bear in mind also that the Black River valley, so open to view from the highway, was not always there, that the Paleozoic section of the Tug Hill which once spanned the gap to the Adirondack foothills has been whittled back by stream and glacial erosion.

Whetstone Gulf State Park, a deep notch cut into the Tug Hill scarp that here exposes Ordovician Whetstone Gulf shale
Courtesy Kendall/Hunt Publishing Co.

Glacial lake deposition also played an important role in the geologic evolution of the modern Black River valley. The east side of the valley south of Port Leyden is distinguished by an exceptionally broad, flat shelf of glacial drift and delta sediments deposited in glacial Lake Port Leyden during Wisconsin deglaciation. The sediment plain is absent from the west side of the valley because the river has removed it as it cut downward and westward down the dip of the Tug Hill strata.

Between Boonville and Alder Creek (7 miles), the route veers to the southeast along a strip of Black River /Trenton limestone, with the Black River about two miles to the east near the Precambrian contact. The headwaters of the river are in Herkimer County in the southwestern Adirondack foothills, narrowly separated from West Canada Creek, which is tributary to the Mohawk River.

South of the highway about two miles northwest of Alder Creek is a pitted outwash plain that contains Echo Lake, a large kettle lake. It marks a stand of the Black River tongue of the receding Wisconsin ice front right at the southern end of the valley.

See map page 358.

NY 46:
Boonville—Rome
24 mi./38 km.

THE BOONVILLE GORGE

This road follows the route of the abandoned Black River Canal which once provided a connecting link between the Mohawk and Black rivers. Several stone locks by the roadside are constructed of hewn blocks of Black River limestone. Near Pixley Falls State Park is the famous "five combines," a flight of five locks by which the canal mounted its steepest ascent. The falls in the park, which cascade over a 50-foot step in the Trenton limestone, mark the break in slope.

The walls of the gorge close in quickly south of Boonville. The narrowing is believed to be at the headwaters divide of a preglacial, north-flowing river that was tributary to the ancestral Black River. Apparently the earlier stream canyon provided a convenient spill-over channel for waters of glacial Lake Port Leyden entrapped in the Black River valley in front of the retreating ice. The torrents of meltwater which emptied into the Lake Oneida arm of Lake Iroquois, cascaded over the divide and eventually whittled it down, so the present gorge has a continuous south gradient from Boonville to Rome. Tiny Lansing Kill (creek), which flows through the gorge and heads at the northern end of it south of Boonville, is now the headwater stream of the Mohawk River, passing through Delta Reservoir en route.

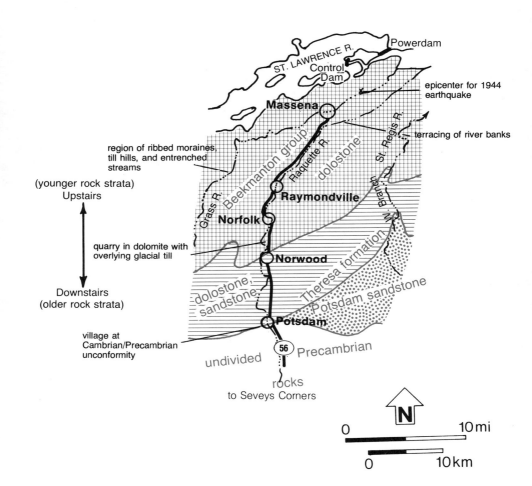

region of ribbed moraines, till hills, and entrenched streams

(younger rock strata)
Upstairs

quarry in dolomite with overlying glacial till

Downstairs
(older rock strata)

village at Cambrian/Precambrian unconformity

epicenter for 1944 earthquake

terracing of river banks

Powerdam

ST. LAWRENCE R.

Control Dam

Massena

Beekmanton group

Grass R.

Raquette R.

dolostone

St. Regis R.

N. Branch

Raymondville

Norfolk

Norwood

Theresa formation

dolostone, sandstone

Potsdam sandstone

Potsdam

56

undivided Precambrian rocks

to Seveys Corners

0   10 mi

0   10 km

NY 56
MASSENA—SEVEYS CORNERS
Also see map page 348.

NY 56:
Massena—Seveys Corners
51 mi./83 km.

The Massena-Cornwall region is of special interest to geologists because a major, magnitude 5.6, earthquake occurred here on September 5, 1944, that did more than $2 million worth of damage (the 1964 "Good Friday" earthquake in Alaska was 8.6). The epicenter, or point on the ground above the earthquake, was at Massena Center, about 3 miles east-northeast of the city. Many smaller earthquakes have been detected in this region before and since the 1944 event; all are undoubtedly caused by fault movement at depth but geologists are still uncertain as to where and how. One interesting observation made by researchers following the 1944 event had to do with the movement of cemetery monuments that were made up of unmortared stones stacked on top of each other. A study of numerous cemeteries on either side of a 20-mile stretch of river revealed a rather remarkable fact: most of the monuments on the American side rotated clockwise and/or shifted one way, while most on the Canadian side rotated counterclockwise and/or shifted in the opposite direction. This at least suggests a fault line parallel to the river along which the Canadian and American ground moved in opposing directions.

The high earthquake potential in this region was, of course, of great concern to builders of the St. Lawrence Seaway and Power Project, a cooperative effort of the United States and Canada that was completed in 1954 (see NY 37: Malone—Ogdensburg discussion). The Moses—Saunders Power Dam and Long Sault Spillway Dam were built across the International Rapids section of the river near Massena, and it was essential to "earthquake-engineer" these structures, as well as the Seaway and locks.

Flooding of the river behind the dams has profoundly changed the shoreline between Ogdensburg and Massena (see NY 37: Malone—Ogdensburg). At Ogdensburg, the effects are minimal, owing to the higher elevation of that part of the river. Changes in morphology increase in magnitude toward Massena, where the river is now 9 times as wide as it was before. One change of geological interest is the partial submergence of numerous "ribbed" moraines to form new islands. The moraines are elongated parallel to the river—and the former Wisconsin ice front—but they have crests with transverse ribs, or serrations that are too subtle to be easily detected on the ground. The ribbing is now exceptionally well defined by the saw-teethed shorelines of some of the new islands, and in others that are more deeply submerged by an east-west string of small north-south-trending islands that resemble a piano keyboard.

Between Massena and Norfolk (9 miles), there are several small, meandering streams tributary to the Raquette River, that have cut deeply into the glacial deposits. Note, for example, how entrenched the Raquette is where the road crosses over it at Raymondville. The Raquette nearer Massena has well-developed terraces on its sides, where it meandered and cut sideways as it also cut downward.

Between Norfolk and Norwood (3 miles), you pass a large stone quarry in the Ogdensburg dolostone, which belongs to the middle Ordovician Beekmantown group. The dolostone bedrock has been quarried primarily for use as an aggregate in concrete or asphalt and for general engineering application in fills, drains, etc. The very finely ground rock has also been sold as agricultural lime.

In the normal operation of such quarries, rock is first blasted from the working face of the pit. The broken rock is then trucked to the top of the pit, where it is passed through a number of crushers and sieves, washed, size-sorted, and stockpiled. Operations like this involving mineral resources of low intrinsic value, are of paramount importance to modern development. This quarry, for example, supplied nearly all of the asphalt aggregate used in road surfacing within a 30-mile radius. It could not be trucked much farther than that because the transportation costs would be too high. Such mineral resources are said to have high place value—they must be close to where they are needed.

Glacial till around the rim of the pit rests directly on the dolostone beds. The erosional surface upon which the till rests is an unconformity, a gap in the geologic record representing about 450 million years of missing rock record.

The village of Potsdam straddles the boundary between the stratified early Paleozoic sedimentary rocks and the Precambrian basement rocks of the Adirondacks upon which they rest. This is also the physiographic boundary between the St. Lawrence Lowlands and Adirondack Mountains provinces. Travel from Massena (18 miles) is downward in the rock section, from the younger Ogdensburg dolostone to older Theresa sandy dolostone. This trend indicates, without question, that the beds dip gently northward away from the Adirondack dome, even though the dip angles are too slight to be detected in single outcrops. Since nearly all sedimentary strata form in horizontal position, their present tilting, all around the Adirondack dome, shows that the uplift has pushed these beds up.

Potsdam is the namesake for the Potsdam sandstone, despite the fact that this formation, which normally separates the Theresa from the Precambrian in this region, is virtually unexposed in the village. The type locality, or exposure for which the formation was first defined, is a few miles to the south at Hannawa Falls. Ironically, the exposures there are atypical of the unit as a whole, and besides, they are part of a small erosional outlier completely isolated from the continuous, wide, outcrop band of Potsdam that borders the Adirondack foothills east of here.

In the past, the Potsdam sandstone was an extremely popular building stone and many of the old buildings in this area are made of it. It is an attractive stone, commonly having red coloration with laminations; it may also be brown, gray, or even white. It splits easily along flat bedding surfaces, facilitating the quarryman's job (see map for NY 3: Watertown—Tupper Lake for the remainder of this route).

Three miles south of Potsdam, near Hannawa Falls, you pass a large sand and gravel quarry pit on the east side of the road. The deposit is in a large kame field, composed of sand and gravel left in deltas along the ice margin by meltwater streams that flowed in or on the ice. Such water-washed deposits tend to be poorly sorted, but moderately well-stratified, sandy, and relatively free of the fine glacial flour found in abundance in glacial till, a direct ice deposit. Like the Norwood stone quarry mentioned above, deposits like this are essential for economic, physical and social growth of a region, and they have a high place value.

Three miles south of Hannawa Falls, just past the Brown's Bridge—Pierrepont road, the road rises on Jenner Hill. The view to the north from the crest of the hill is one of the best panoramas in the entire North Country of the St. Lawrence Lowlands. When approached from the south, with its rolling foothills topography, this sudden apparition makes the descent from the mountains a vivid reality.

View north to the St. Lawrence lowlands from Jenner Hill

Much of the foothills country bordering the St. Lawrence Lowlands is practically buried in sand, as you see along NY 56 between Hannawa Falls and Seveys Corners (29 miles). Most of the deposits are of simple deltaic origin; they formed at the mouths of streams like the Raquette or St. Regis rivers that flowed from the Adirondacks into Lake Iroquois, the large proglacial predecessor of Lake Ontario. Different delta levels reflect different surface levels of Lake Iroquois. Near Hannawa Falls, for example, is a delta plain at 650 feet above sea level, while another occurs between Hannawa Falls and Colton at

Parishville "desert" on deltaic sand and gravel deposits formed in glacial Lake Iroquois; boulders form a pavement as a result of wind evacuation of sand.

860 feet. Some deposits formed as valley-filling deltas in smaller meltwater lakes within the foothills. Many of the cleaner, sandier deposits are actually beach-and-dune sediments. Wind erosion must have been extremely effective in that immediate postglacial period, when the land was barren, covered with unconsolidated sediments and unshielded by a protective blanket of vegetation.

The bedrock boundary between the Lowlands and Highlands Adirondacks subprovinces is passed at Colton. On a regional scale, the boundary separates the dominantly metasedimentary/ metavolcanic rocks of the Lowlands from the dominantly metamorphosed igneous intrusive rocks of the Highlands Adirondacks. This boundary winds back and forth for 70 miles in a northeasterly direction between Carthage and Parishville, beyond each of which it is concealed under alluvium or Paleozoic strata. Downstream from the Colton dam, the Raquette River has cut a spectacular little gorge called Stone Valley through the boundary zone, where its character is revealed. The boundary is a fault zone marked by intense deformation of a complex array of interleaved and infolded gneisses, amphibolites, quartzites, calc-silicate rocks, and marbles. The rocks are solid and their deformation appears to be well-annealed by metamorphic recrystallization. The valley's most striking aspect, exposed by river erosion, is the planar cleavage of the bedrock that dips step-like in one direction and results from fault-induced shearing.

The dams at Hannawa Falls and Colton are just two of the 19 on the Raquette River between Tupper Lake and the St. Lawrence River, a riverbed distance of about 100 miles. The hydroelectric capacity of the river is more extensively developed than almost any other small river in the eastern United States. The river has even more potential which will probably never be realized because most of it lies within the Adirondack Park, where development is now severely restricted.

Between 4 and 6 miles south of the South Colton dam, you follow a straight, level stretch of highway over "the Plain," a remarkable example of a meltwater lake basin that was filled with sediments, mostly of deltaic origin. These are extremely common features of the Highland Adirondacks. During and after Wisconsin deglaciation, there were many hundreds more lakes than there are today. Most were rapidly filled with sediments transported from the barren landscape by inflowing streams. Some remain today, only partially filled and boggy, like the several by the highway between the Plain and Seveys Corners (12 miles).

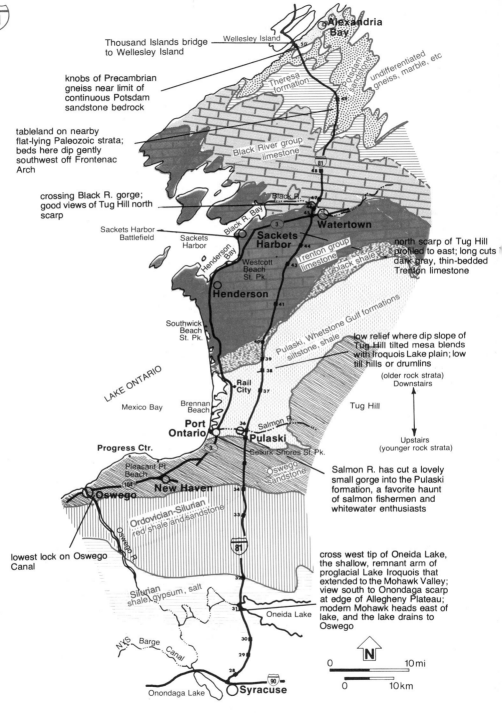

Thousand Islands bridge to Wellesley Island

knobs of Precambrian gneiss near limit of continuous Potsdam sandstone bedrock

tableland on nearby flat-lying Paleozoic strata; beds here dip gently southwest off Frontenac Arch

crossing Black R. gorge; good views of Tug Hill north scarp

Sackets Harbor Battlefield

north scarp of Tug Hill profiled to east; long cuts dark gray, thin-bedded Trenton limestone

low relief where dip slope of Tug Hill tilted mesa blends with Iroquois Lake plain; low till hills or drumlins

(older rock strata) Downstairs

Tug Hill

Upstairs (younger rock strata)

Salmon R. has cut a lovely small gorge into the Pulaski formation, a favorite haunt of salmon fishermen and whitewater enthusiasts

lowest lock on Oswego Canal

cross west tip of Oneida Lake, the shallow, remnant arm of proglacial Lake Iroquois that extended to the Mohawk Valley; view south to Onondaga scarp at edge of Allegheny Plateau; modern Mohawk heads east of lake, and the lake drains to Oswego

Alexandria Bay
Wellesley Island
Theresa formation
Potsdam sandstone
undifferentiated gneiss, marble, etc

Black River group limestone

Black R.

Watertown

Sackets Harbor
Sackets Harbor
Black R. Bay
Henderson Bay
Trenton group limestone
black shale

Westcott Beach St. Pk.

Henderson

Southwick Beach St. Pk.

Pulaski, Whetstone Gulf formations siltstone, shale

LAKE ONTARIO

Rail City

Mexico Bay

Brennan Beach

Port Ontario

Salmon R.

Pulaski

Progress Ctr.

Selkirk Shores St. Pk.

Pleasant Pt. Beach

Oswego sandstone

Oswego **New Haven**

Oswego R.

Ordovician-Silurian red shale and sandstone

Oneida Lake

Silurian shale, gypsum, salt

N.Y.S. Barge Canal

Onondaga Lake

Syracuse

N

0   10 mi
0   10 km

Interstate 81
SYRACUSE—ALEXANDRIA BAY

370

Interstate 81:
Syracuse—Alexandria Bay
95 mi./154 km.

Between Syracuse and Watertown (70 miles), Interstate 81 skirts the low dip-slope side of the Tug Hill (see NY 12: Watertown—Boonville discussion), where it blends with the Ontario Lowlands. The route, therefore, is rather flat, and there are virtually no road-cuts except near Pulaski and Watertown. The great drumlin field of western New York begins in earnest just west of Syracuse, with most of the drumlins occurring between there and Rochester. The well-formed drumlins do, however, continue northward to beyond Watertown in decreasing numbers, but most of them lie west of the highway near the Ontario shores.

The most prominent topographic feature as you leave Syracuse is the Onondaga scarp slightly south of the city. The east-west scarp marks the boundary between the Ontario Lowlands and the Allegheny Plateau physiographic-geologic provinces, and it is held up by the resistant Onondaga limestone near the base of the Devonian strata of western New York. The dominant topography of the plateau region is that of a tilted mesa, with cliffs facing northward and gentle dip slopes southward (see US 11: Potsdam—Watertown discussion). Here, and for some distance east and west, the Onondaga is the most prominent cliff-former; farther east it is the underlying Helderberg group limestones.

At mile 100, you cross the Oneida Lake bridge. The lake is really just a shallow remnant of an arm of glacial Lake Iroquois that extended eastward to the Mohawk Valley. The Mohawk at that time, before the Wisconsin ice sheet receded far enough northward to open the St. Lawrence River outlet, carried nearly all of the outflow from the newly-forming Great Lakes. Oneida Lake now drains to Lake Ontario by way of the Oswego River at its west end, and the Mohawk begins north of Rome, 16 miles east of the lake on the other side of a drainage divide.

Drumlins in the region immediately north of Oneida Lake are elongated in a northwesterly direction with their blunt, "upstream" ends toward Lake Ontario. They radiate from the lake. In fact, there is a persistent radial arrangement around the lake throughout the drumlin field, from Niagara Falls all the way to Watertown, indicating that the lake basin acted as a spreading center for the Wisconsin ice sheet. This is further substantiated by similar radial trends of glacial striae or scratches where the ice scraped against bedrock.

Most of the relief along the route from Syracuse at least to Adams Center, 14 miles south of Watertown, is depositional-glacial, glacial-river, and glacial-lake sediments left on nearly flat-lying Silurian and Ordovician beds. As such, the topography is not very different from that of the St. Lawrence Lowlands, with a gently undulating surface and a lot of marshy bottomland. The road climbs over a few drumlins and cuts through several. The concealed boundary between Silurian and Ordovician rocks is passed at the Mexico exit (27 miles from Syracuse); from here north to Pulaski, the bedrock is the Oswego sandstone, the caprock of the Tug Hill tilted mesa.

At Pulaski, you pass over the Salmon River, which has cut a small gorge into the interlayered siltstones and shales of the Pulaski formation. The river is both a salmon fisherman's haven and a whitewater favorite in spring when the runoff is high. Many small streams that cross this route are slightly entrenched for much the same reasons that streams in the St. Lawrence Lowlands are entrenched (see NY 37: Malone—Ogdensburg discussion), but the downcutting is not as great because the glacial rebound of the land has been less.

Near Watertown, the topography of the northern edge of the Tug Hill is well profiled east of the highway, with north-facing scarps, and gentle south dip slopes built on bedding surfaces. Several tilted mesas are to be seen here with Trenton limestone caprock, all of which are erosional remnants of the main Tug Hill plateau and that form stair-steps that descend northward. The highest step is Dry Hill, visible east of the road over most of the 8 miles from Adams Center to Watertown. The caprock surface, which slopes slightly toward you as well as southward, forms a scarp about 300 feet high at the northern end. Dry Hill is separated from the main Tug Hill uplands by Sand Creek on its eastern side. Watertown's Thompson Park southeast of the city is atop a much smaller tilted mesa separated from Dry Hill by Mill Creek.

Between Exits 44 and 45, you pass long cuts in thin-bedded, dark gray Trenton limestones as the road descends to Beaver Meadows through a small scarp. At Exit 45 on the north side of the meadows, the same beds are exposed along the access road for the southbound lane. Between Exits 46 and 47, the highway passes over Black River, with broad banks of massively-bedded, light gray Black River limestones. Numerous small caves have openings in rocky banks farther downstream. The river issues from the Black River Valley east of the Tug Hill, and wraps around its northern end on the way to Lake Ontario west of Watertown, close to the contact between the Black River limestone and Trenton limestones. The Black River limestone here may be reocgnized by its gray color and generally thick beds, whereas the Theresa is somewhat darker grayish-brown and relatively thin-bedded. The Potsdam has a distinctive tan to pinkish-tan color in this region and is evenly bedded.

Between Exit 48 and the Thousand Islands Bridge south of Alexandria Bay, the road traverses nearly level land developed over, first Black River limestone, then Theresa formation and finally Potsdam sandstone, not far from the contact between the latter formation and the underlying Precambrian gneisses, marbles, and schists of the Lowlands Adirondacks. The lower Paleozoic beds here dip slightly to the southwest as a result of the geologically young rise of the Frontenac Arch. The arch trends northwestward across the St. Lawrence River and is responsible for the erosional stripping of the

Black River from Interstate 81 bridge at Watertown, with Black River limestone in banks

Theresa and Potsdam beds and exposure of the underlying Precambrian terrane in the Thousand Islands region. Nearing the Thousand Islands Bridge, a number of glacially-polished gneiss knobs are exposed by the highway where they jut through the Potsdam beds. The topography displayed by this tough bedrock is apparently much the same as it was some 525 million years ago, when the Potsdam Sea first invaded the region and the earliest Potsdam sands began to cover it up.

Pothole in gneiss

374

See map page 348.
US 11:
Potsdam—Watertown
70 mi./113 km.

Potsdam straddles the Raquette River right at the boundary between continuous, early Paleozoic layercake sedimentary strata of the St. Lawrence Lowlands and Grenville-age Precambrian metamorphic rocks of the Adirondacks. The village is the namesake for the famous Potsdam sandstone, the attractive reddish, white, or brownish building stone used so widely in this region during the last century. Many of the stately old buildings in the village are constructed of this popular stone. Much sandstone for buildings throughout the area was taken from quarries on both sides of the Raquette River south of Potsdam. Ironically, the quarries lie in an erosional outlier, resting on, and completely surrounded by, Precambrian rocks; there are no significant exposures in Potsdam itself.

Between Potsdam and Canton, the bedrock is all Precambrian gneiss and marble, but most of it is concealed beneath unconsolidated glacial till and soil. The road goes up and down over gently undulating, drumlin topography, with a lot of marshy bottomland, conditions typical of much of Lowlands Adirondacks and St. Lawrence Lowlands. Near Canton, a high section of the road affords a panoramic view southward to the moderately corrugated profile of the Adirondack foothills in the distance. Drumlins here tend to be of the classic, inverted-spoon form with north-south elongation and steep north ends. But they are not nearly as well developed nor as large as those of the great drumlin field between Syracuse and Rochester.

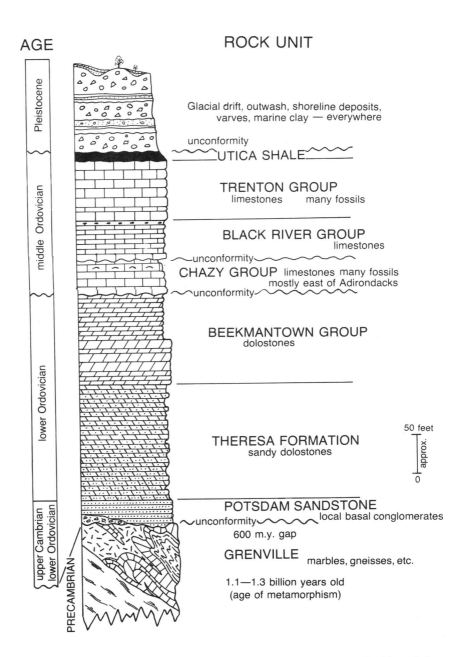

AGE

ROCK UNIT

Pleistocene

Glacial drift, outwash, shoreline deposits, varves, marine clay — everywhere

unconformity

UTICA SHALE

TRENTON GROUP
limestones   many fossils

BLACK RIVER GROUP
limestones

unconformity

CHAZY GROUP limestones many fossils
mostly east of Adirondacks

unconformity

BEEKMANTOWN GROUP
dolostones

THERESA FORMATION
sandy dolostones

50 feet

approx.

0

POTSDAM SANDSTONE
local basal conglomerates

unconformity

600 m.y. gap

GRENVILLE marbles, gneisses, etc.

1.1—1.3 billion years old
(age of metamorphism)

middle Ordovician

lower Ordovician

upper Cambrian / lower Ordovician

PRECAMBRIAN

Composite stratigraphic section for the northwest and the north sides of the Adirondacks, straddling the Frontenac arch. The rocks are stacked in chronological order with the oldest at the bottom; glacial sediments are everywhere and cover rocks of all ages.
Courtesy Kendall / Hunt Publishing Co.

In Canton, there are two bridges over the Grass River where Pre-cambrian gneisses may be seen both upstream and downstream. These rocks lie at the edge of the Canton "alaskite" body, the largest of the numerous masses of pinkish, granitic gneisses now referred to collectively as the Lower gneiss. This is one of the three major rock types of the Lowlands Adirondacks, the others being the Upper gneiss and Gouverneur marble. These three, broadly defined units are now thought to be interlayered and complexly folded and faulted on a grand scale to produce the intricate, fluid patterns visible on geologic maps. Exposures of dark gneiss with pink granitic layers and lenses appear at the railroad underpass just south of Canton.

Migmatite of the "upper gneiss," containing dark gneiss with lenses, layers, and interfingering of light-colored granite.
Courtesy Kendall/ Hunt Publishing Co.

Between Canton and DeKalb Juncton, several smoothly rounded "gneiss knuckles" that are characteristic of the Lowlands Adirondacks landscape project from hayfields. Really elongate low ridges that trend northeast-southwest, these hard rock barriers have exerted strong control over drainage trends. Major rivers, like the Oswegatchie, that drain across this region from the highlands, are forced to zig and zag between the ridges on their way to the St.

Lawrence; tributaries to them are channeled in a similar manner. Many of the intervening valleys are underlain by marble, which is less resistant than the gneisses and is therefore poorly exposed. In addition to drainage control, this northeast structural "grain" has influenced human development; US 11, for example, runs parallel to it because it is easier and cheaper to build a road that way. The route is part of the Old Military Turnpike discussed in more detail in the Potsdam—Plattsburgh section.

There are many excellent roadcuts south of Canton. The large "Snake" roadcut, in whitish marble 4 miles from Canton, shows, on the southeast side, a thin interlayer of feldspathic rock that has been sinuously infolded with the marble, so that it looks like its namesake. The Snake probably originated as a flat-lying layer of volcanic ash deposited with the parent limestone of the marble, in a marine environment that existed perhaps 2-3 billion years ago. Subsequent Grenville metamorphism, 1.1 billion years ago, converted the rocks to their present crystalline state and caused the intense deformation shown here. The Snake effectively delineates the folding style of the

The "Snake" roadcut, 4.2 miles southwest of Canton on US 11, an example of highly deformed Gouverneur marble which behaved plastically during Grenville metamorphism; the Snake is a band of whitish feldspar (here marked on the photo with black ink), apparently formed by metamorphism of a layer of volcanic material sandwiched between the original limestone beds.

marbles, which is decidedly plastic as opposed to brittle. Adirondack marbles like this often contain bands of thinly interlayered silicate rocks that illustrate just how complex deformation can be. The silicate layers weather in relief and are usually rusty, or they contrast sharply with the whitish, recessively-weathered marble in between.

Several more excellent roadcuts are between DeKalb Junction and Richville (9 miles). One of the most dramatic is the Red and White cut 4 miles south of DeKalb Junction. Really comprising several cuts on both sides of the road on a long, gradual hill, the white is Precambrian marble, and the red is iron-stained conglomerate and sandstone, all assigned to the basal Potsdam sandstone. Contacts between the marbles and the sedimentary materials are irregular, the sediments have a dumped appearance, and there are blocks of marble in them. These and other features leave little doubt that the sandstone is one of the sinkhole fillings so common in the marbles of the Lowlands Adirondacks. Apparently most of the Precambrian bedrock of the Adirondack region was eventually blanketed with sandy and conglomeratic sediments during its immersion in the Potsdam Sea; thick pockets occur wherever a solutional topography of sinkholes and caves had previously developed in marbles. Doming of the Adirondacks has since stripped the sandstone cover, except in pockets like this near the St. Lawrence Lowlands.

Sinkhole-filling of Cambrian Potsdam sandstone in Precambrian marble with inclosed blocks of marble; Rock Island Road

Cylindrical structure in Potsdam sandstone, presumably formed where sand grains were arranged by sifting down into a small sinkhole in the underlying Precambrian marble, like an hourglass—Cream of the Valley Road.
Courtesy Kendall / Hunt Publishing Co.

Roadcuts at the south end of the Richville bypass, 10 miles south of DeKalb Junction display similar features, with brick red conglomeratic rocks on the west and marbles on the east. These and other cuts a short distance south of the bypass also contain reddish conglomeratic dikes, where the original, water-saturated sediments filled narrow fractures in marble.

Between Richville and Gouverneur, the gneiss ridge-marble valley topography is especially well-developed. In Gouverneur, note the several buildings constructed of gray, rough-hewn, Gouverneur marble, an extremely popular stone and architectural style throughout the North Country during the 19th century. In some buildings, Potsdam sandstone is used as decorative trim, giving a striking gray-red pattern to the stonework.

In Gouverneur, you pass over the Oswegatchie River, which follows a remarkable course downstream, its path involving a series of hairpins controlled by the distribution of hard and soft rock. The river has been forced to follow a tortuous 27-mile course to advance just 4 miles northwestward toward the St. Lawrence River!

Between Gouverneur and Antwerp (12 miles), there are many clean roadcuts in a wide variety of Precambrian rock types, but whitish banded marble is the most common. Seven miles south of Gouverneur, an erosional outlier of Potsdam sandstone occurs on the west side. At the top of this cut is a bowl-shaped structure in the sandstone of the type that is found in many North Country exposures of the Potsdam on both sides of the St. Lawrence River. These peculiar structures appear to be located close to the base of the formation where it rests on marble. They are thought to form as the original water-saturated sand sifted into small solutional pockets in the underlying marble, rather like an hourglass. Hundreds of these structures of all sizes are exposed in pastureland along the Cream of the Valley road 8 miles north of Gouverneur. The smaller ones tend to be more cylindrical; the larger ones are more bowl-shaped, more irregular, and more broken up by pre-consolidation deformation.

A sand and gravel pit on the west, 8 miles south of Gouverneur, is in a glacial kame. Kame deposits form when meltwater streams dump their sediment load into a lake at the edge of the glacier. Because they are stream-carried, the deposits are generally sandy and well stratified, with more or less well-rounded gravel.

Fluid deformation of marble (upper), as compared to brittle response of adjacent gneiss (lower)

Several of the marble cuts between Gouverneur and Antwerp (13 miles) incorporate dark greenish-gray calc-silicate rocks, made up of calcium- and silicate-bearing minerals, in the form of bands of broken, displaced blocks. They illustrate how different rock types react to stress; deformational patterns in the marble are quite fluid, or taffy-like, while those of the calc-silicate are brittle, like the snapping of a piece of blackboard chalk. These rocks were, of course, deformed far below the surface where temperature and pressure were extreme.

"Train wreck" of rectangular amphibolite blocks (dark), possibly from basaltic dike intruded into marble (or limestone) prior to final deformation and metamorphism; illustrates brittle reaction of dark blocks as compared to plastic reaction of marble.

Between Antwerp and Philadelphia (5 miles), the road parallels the Indian River which, like the Oswegatchie, follows a devious northeast-southwest course that periodically crosses the hard ridges and travels many extra miles to get to the St. Lawrence River. Continuing nearly 8 miles to Evans Mills, the river then makes a sharp turn and goes 8 miles straight north to Theresa as it cuts through nearly flat-lying Potsdam sandstone and Theresa formation near the Precambrian boundary. The river there established its course on the Paleozoic strata independent of structural channelways in the underlying metamorphic rocks, and, locally, it cut down into them without changing direction. Such a river, geologists say, is "superposed."

South of Philadelphia, the road follows a ridge of Potsdam sandstone and Theresa formation with good views of the distant northern scarp of Tug Hill. The broad, gentle slopes and low relief reflect the character of the bedrock. At Evans Mills, the road climbs over a ridge of Black River limestone by the water tower. A quarry in this unit is on the south side of the hill.

Most of the way between Evans Mills and Watertown (9 miles), the route traverses a Black River limestone shelf that borders the Tug

382

Hill scarp. Tug Hill is rather similar to the flat-topped mesas of the southwestern United States, except that the strata dip, or slope, gently in one direction, generally westward here. Long term erosion produces a steep scarp on the upturned end, and a gentle dip slope on the downturned side, in other words, a tilted mesa. The hard layer or caprock at the top of Tug Hill is the Ordovician Oswego sandstone. Resistant lower strata, like the Black River and Trenton limestones, form stair-steps in front of the main scarp. What you see, therefore, from north of Watertown along US 11, is not just one great scarp, but several erosionally-isolated, small tilted mesas.

Abandoned Black River limestone quarry at Evans Mills

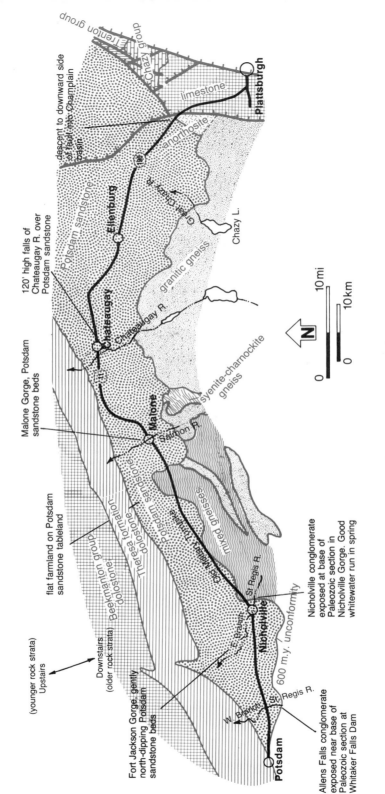

The Old Military Turnpike
POTSDAM—PLATTSBURGH

Route 11
Route 190

Plattsburgh

Trenton group

Chazy group

Chazy group

limestone

descent to downward side
of fault into Champlain
basin

anorthosite

Potsdam sandstone

Ellenburg

Route 190

Great Chazy R.

Chazy L.

granitic gneiss

120' high falls of
Chateaugay R. over
Potsdam sandstone

Chateaugay

Chateaugay R.

Route 11

Malone Gorge, Potsdam
sandstone beds

Malone

syenite-charnockite
gneiss

Salmon R.

flat farmland on Potsdam
sandstone tableland

Beekmantown group
dolostone

Theresa formation
dolostone

Potsdam sandstone

Old Military Turnpike

mixed gneisses

(younger rock strata)
Upstairs

Downstairs
(older rock strata)

Nicholville conglomerate
exposed at base of
Paleozoic section in
Nicholville Gorge. Good
whitewater run in spring

St. Regis R.

Nicholville

E. Branch

600 m.y. unconformity

Fort Jackson Gorge; gently
north-dipping Potsdam
sandstone beds

W. Branch

St. Regis R.

Potsdam

Allens Falls conglomerate
exposed near base of
Paleozoic section at
Whitaker Falls Dam

N

0   10 mi
0   10 km

THE OLD MILITARY TURNPIKE
FROM POTSDAM TO PLATTSBURGH

This North Country route follows the Old Military Turnpike that was used extensively during the War of 1812 for the transport of troops, munitions, and even boats between Sackets Harbor and Plattsburgh. The routing of the turnpike demonstrates how human activity is often moderated or controlled by geology. Nearly the entire route lies atop Potsdam sandstone just north of its boundary with the complex metamorphic rocks of the Adirondacks. Relief along the route is moderate, and the soil is generally sandy and well drained, highly suited to farming and roadbuilding. By contrast, the region immediately south is rocky, hilly, densely forested, poorly drained, and dotted with lakes, ponds, and bogs. The route, therefore, presented the American patriots with the shortest and easiest east-west traverse around the north side of the Adirondack barrier.

Potsdam is the namesake for the famous Potsdam sandstone, but there are virtually no outcrops in the village of this widely-used building stone. Five miles east of the village is the high bridge over the west branch St. Regis River, a favorite among whitewater enthusiasts in April and May when the river is high. A few miles upstream, the Allens Falls conglomerate is exposed at the base of the Potsdam sandstone; a little farther up, at Whitaker Falls dam, Precambrian gneisses form the streambed, and basal Potsdam conglomerates are exposed in the banks. Six miles downstream from the bridge, at West Stockholm, the stream passes over stepped beds in the Theresa formation. Similar younging, i.e., traversing from older to younger sedimentary layers, occurs along all the major streams that drain northward from this route. The northward dip of the beds, though gentle, is steeper than the gradients of the streams that cut across them. This dip away from the Adirondacks suggests that Paleozoic strata may once have covered the Precambrian rocks completely, but have been eroded away as a result of the geologically recent, and current, doming of the mountain mass. At Southville, the road passes near the concealed boundary between the Lowlands and the Highlands Adirondacks. The lithological change across the boundary is abrupt along its entire length, which winds back and forth from here in a south-southwest direction to beyond Natural Bridge, where the Adirondack rocks are again concealed under Paleozoic strata.

Between Hopkinton and Malone (25 miles), the unobstructed views northward over farmland, to the strikingly flat St. Lawrence Lowlands, give the apt impression, especially on hazy days, of gazing out to a distant sea over an even-graded coastal plain. It is as if the early Paleozoic sea were still there. To the south, the land rises slightly and becomes more rugged over the metamorphic rocks of the Lowlands Adirondacks.

At Nicholville, another high bridge passes over the east branch St. Regis River, which is also a whitewater favorite in spring. A mile upstream, on the Port Kent Road, the river has carved a narrow gorge into the Nicholville conglomerate, exposing Precambrian gneisses in the streambed. Like the Allens Falls conglomerate, this is just one of the numerous units found resting on gneisses at the base of the Potsdam sandstone in many parts of the North Country. At Fort Jackson, three miles downstream from the bridge, another little gorge is cut through more "normal" Potsdam, with pronounced bedding and crossbedding.

Between Nicholville and Bangor, the ground is somewhat hillier as the road repeatedly crosses the Precambrian boundary. Many stony fields and stone fences betray the glacial origin of the surface sediments, but the relief is principally controlled by bedrock resistance, especially just east of Bangor where the road rises about 250 feet.

Between Bangor and Malone, the even, gently-sloping farmland over Potsdam beds contrasts with the Bangor terrain. The Potsdam is exposed in a small canyon of the Salmon River near the center of Malone.

On NY 11, between Malone and Ellenburg (25 miles), the topography is similar, as the route projects farther northward onto the Potsdam surface. Just west of Chateaugay, you pass over the deeply entrenched Chateaugay River. North-flowing streams along this route are all entrenched to some degree, as a result of deglaciation at the end of the Ice Age. While the St. Lawrence Valley was still blocked by ice, Lake Iroquois occupied the Ontario basin and extended far down the St. Lawrence, where its waters lapped high on the slope south of the present river. The erosional base level then for rivers like the Chateaugay was the lake surface. When the St. Lawrence was finally free of ice, Lake Iroquois drained through it to a lower level, and streams thus began rapid down-cutting to form gorges, rapids, and waterfalls. This process was further enhanced by rebound of the land that had been deeply depressed under the ice, a spring-back that is still going on, and which has raised the surface more than 500 feet in this region. Chateaugay Falls, about a mile southwest of Chateaugay, is one of the loveliest of all, in fact, of the entire state. Here, the river cascades, in two steps, over 120 feet of well-bedded Potsdam sandstone, through a narrow, tree-shrouded crevice.

Between Ellenburg and Plattsburgh on NY 190 (25 miles), the Military Turnpike skirts the northeast corner of the Adirondacks and drops into the Champlain Lowlands. In the southern half of the route, the road descends more than 500 feet as it crosses to the downdropped side of one of the major block faults that line the western side of the Champlain basin. The route is punctuated with frequent glimpses of Lake Champlain to the southeast, and roadcuts in whitish Precambrian anorthosite. Near West Beekmantown are the ruins of an old stone house, with an historical marker that reads "1823—The remains of Robinson's Tavern, erected by Lewis Sage Robinson opposite from an original log tavern visited in 1817 by President James Monroe." This is one of the many inns and other establishments that originally dotted the Old Military Turnpike.

Most of the way from the ruins to Plattsburgh, the road is quite level, as it passes over flat-lying Beekmantown dolostone beds.

High falls of Chateaugay River; 150-foot drop over Potsdam sandstone.

GLOSSARY

Adirondack arch: ridge of Precambrian rock exposed in the Thousand Islands

Adirondack dome: overturned-bowl shape of the Precambrian erosional surface in and around the Adirondacks, partly concealed beneath Paleozoic sedimentary rocks, and apparently due to spot-like uplift which continues today

Alaskite: granite with an extremely low proportion of dark minerals

Alluvium: unconsolidated, stream-laid sediments

Amphibolite: dark metamorphic rock chiefly composed of hornblende and plagioclase feldspar

Anorthosite: igneous rock composed of 90% or more plagioclase feldspar; metanorthosite is the metamorphosed equivalent

Anticline: upfolded rock layers

Asthenosphere: the soft layer that supports the hard crust of the earth with its moving tectonic plates

Augen: "eyes," a German term used to describe eye-shaped lenses of minerals, such as in augen gneiss

Badlands: intricate erosional surface in soft materials, with steep, narrow ravines and sharp crests and pinnacles

Barrier island: offshore sand island built by wave action and longshore drift

Basalt: dark, fine-grained volcanic rock

Basement rocks: normally very old rocks upon which younger sediments and/or volcanic materials are deposited

Baymouth bar: ridge of sand that blocks the mouth of a bay, caused by longshore drifting of beach sand

Bed: sedimentay layer with more or less distinct rock-type and upper and lower boundaries

Bedrock: continuous solid rock either exposed at the surface or overlain by unconsolidated sediments; part of the earth's crust

Block faulting: type of faulting by which the crust is divided into large blocks of different elevation by a series of up-down faults, generally with parallel trends

Bog: spongy morass chiefly composed of decaying vegetable matter or peat; a swamp

Calc-silicate rock: metamorphic rock mainly composed of calcium silicate minerals such as hornblende, epidote, and diopside, often with calcite or dolomite

Catskill delta: huge delta complex of middle and late Devonian sedimentary rocks that underlie the Allegheny Plateau of New York State (and beyond)

Champlain Sea: marine invasion of the St. Lawrence and Champlain valleys in the waning stages of Wisconsin glaciation

Charnockite: high grade granitic gneiss that contains hypersthene (a pyroxene)

Chatter marks, glacial: rows of small, cusp-shaped joints in bedrock caused by "chattering" of ice-bound rocks against it as the glacier moves

388

Chill zone: fine-grained border zone in igneous rocks where the melt has cooled quickly against the country rock or by exposure to air or water

Cirque: half-bowl-shaped rock amphitheater formed at the head of a valley, or alpine, glacier

Composite stratigraphic column: graphical composite of two or more rock sections with rock-stratigraphic units "stacked" in their proper order of oldest at the bottom and progressively younger upwards

Conglomerate: coarse-grained sedimentary rock with large, rounded pebbles or boulders in a finer-grained matrix

Continental glacier: moving ice sheet of continental proportions

Continental shelf: gently-sloping, submerged edge of the continent out to the continental slope, with a maximum depth of about 600 feet

Continental slope: outer edge of the continental shelf

Correlation, stratigraphic: determination of time-equivalence between units of geographically separated rock sections

Cross-bedding: sets of inclined sediment or sedimentary rock beds formed in dunes, deltas, or beaches, especially in sand

Crystal: geometric form of a mineral with plane faces that are the external expression of an internal atomic order

Crystalline rock: igneous and metamorphic rocks

Cuesta, or tilted mesa: assymetric ridge wth a scarp face and a gentle "dip slope" developed on top of resistant beds

Decomposition, rock: chemical weathering of rock, as in solution of limestone

Deglaciation: uncovering of glaciated land by ice wastage

Delta: generally fan-shaped body of sediment formed at the mouth of a river

Diabase: basaltic rock chiefly composed of small, well-formed plagioclase crystals in a dark matrix of pyroxene

Dike: tabular intrusion of magma or (less commonly) water-saturated sediments into cross-cutting fractures

Dip: angle of slope of rock layers as measured from the horizontal

Discharge, river: rate of flow through a certain cross-section of a river over a specific time, often expressed as cubic feet per second (cfs)

Disintegration: mechanical weathering of rock

Dolostone: sedimentary rock chiefly composed of the mineral dolomite

Drift, glacial: all sediments of glacial origin

Dripstone: general term for cave deposits formed by dripping water

Drumlin: till hill streamlined by overriding ice, with an ideal shape of an overturned spoon

Earthquake: sudden trembling of the land caused by abrupt release of accumulated strain along a fault or in an erupting volcano

Epoch: geologic time unit next smaller in order to period

Era: geologic time unit including several periods

Erie Canal: the most famous of New York's barge canals that were developed for commerce and transportation in the 19th century, completed in 1825, connecting the Hudson River at Albany with Lake Erie at Buffalo

Erosion: sum of all geologic processes that tend to wear down the land

Erratic, glacial: boulder transported by a glacier that generally differs from the bedrock underneath

Esker: long, low, sinuous ridge of sand and gravel dropped by streams that flowed on or through tunnels in a retreating glacier

Extrusive igneous rock: volcanic rock

Fault: surface or zone of rock fracture caused by bodily movement of one mass of rock against another

Fault scarp: cliff on the upthrown side of an up-down fault

Flour, glacial: pulverized rock produced by grinding together of rocks in and under glacial ice

Flowstone: general term for carbonate deposits formed on cave walls or floor by flowing, rather than dripping, water

Fold: rock deformation manifested by real or apparent bending of bedding, foliation, or layering

Foliation: general term for planar textural or structural features in a rock, most often applied to metamorphic rocks

Formation (stratigraphic): basic, mappable, rock-stratigraphic unit that is more or less homogeneous

Fossil: any remains, trace, or imprint of plant or animal naturally preserved in sediments or rocks from past geologic time

Fossil soil: buried and preseved soil horizon from the geologic past

Frost action: mechanical weathering caused by alternate freezing and thawing

Frost-heave: expansion and upward movement of water-soaked soil when it freezes

Gabbro: coarse-grained, dark igneous intrusive rock mainly composed of plagioclase feldspar and pyroxene; intrusive equivalent of basalt

Geosyncline: large, linear, downwarping segment of earth crust that fills up with sediments to tens of thousands of feet, as along continental margins

Glen: deep, narrow canyon cut by a stream at the site of a hanging valley, also called gulf, hollow, or clove

Gneiss: coarse-grained, foliated metamorphic rock

Graben: in block-faulting, an elongate, down-dropped crustal block forming a valley, typically bounded by parallel, inward-dipping faults (see also Horst)

Granite: coarse-grained igneous intrusive rock chiefly composed of potassium feldspar and quartz

Granite veining: lenses and layers of granite found in some gneisses (migmatites)

Graywacke: "dirty," grayish sandstone containing much clay, feldspar, and angular rock fragments

Grenville mountain-building event: affected the east side of ancestral North America about a billion years ago as the continent collided with ancestral Africa

Grooves, glacial: large, linear furrows caused by scraping of ice-bound rocks against bedrock as the glacier moves

Groundwater: water held in rocks and sediments in pore spaces or other openings

Group, stratigraphic: major rock-stratigraphic unit next higher in rank than formation, consisting wholly of two or more contiguous formations

Gulf: see Glen

Hanging delta: delta formed where a glacial meltwater stream empties into a lake that is now gone or diminished, leaving the deposits suspended from the valley wall

Hanging valley: small glacial valley that joins a large trunk glacial valley high above its floor

Horst: in block-faulting, an elongate, elevated block of crust forming a ridge or plateau, typically bounded by parallel, outward-dipping faults (see also Graben)

Ice sheet: thick, unconfined, sheet-like glacier, generally covering a very large area, that spreads radially over the land under its own weight

Igneous rock: formed by solidification of rock melt or magma

Insurgence: point at which surface water disappears into a cave system and becomes an underground stream

Interglacial: time period between major glaciations of the Pleistocene Ice Age

Intrusive rock: formed from magma that crystallized below the surface

Iroquois, Lake: proglacial lake that occupied the Lake Ontario basin in the waning stages of Wisconsin glaciation, extending well beyond the present lake shores

Joint: rock fracture that simply opens up by tension, and the opposing blocks do not slide past each other

Kame: generally conical hill of sand and gravel deposited in a proglacial or englacial lake by meltwater streams

Kame terrace: terraced kame deposits on valley sides formed as a glacier tongue recedes upvalley while the downvalley section contains a proglacial lake

Karst: irregular, pitted topography characterized by caves, sinkholes, disappearing streams, and springs, and caused by water solution of underlying limestone, dolomite, or marble

Kettle: depression in glacial deposits where outwash was deposited around a residual block of ice that later melted

Klippe: isolated erosional remnant of an overthrust low-angle push fault

Lava: rock melt or magma that reaches the surface; also the same material solidified by cooling

Limestone: sedimentary rock chiefly composed of the mineral calcite

Lobation, glacial: division of an ice sheet into lobes that reflect the underlying valleys

Magma: rock melt that may or may not reach the surface and forms igneous rocks when it cools and crystallizes

Mantle, earth: concentric, interior zone of earth below the crust and above the core, ranging from depths of about 40 to 3480 kilometers (24 to 2088 miles)

Marble: metamorphosed limestone or dolostone

Meanders: broad, semi-circular curves in a stream course that develop as the stream erodes its outer bank and deposits sediment against the inner bank

Melange: mappable mixture of deformed rocks formed at the base of a thrust fault, often with fragments derived from formations of widely different ages

Meltwater: water from the melting of ice and snow, especially in glaciers

Member: rock-stratigraphic subdivision of formation, and consisting of many beds

Meta- (prefix): signifies that the rock has been metamorphosed, as in metanorthosite, metagabbro, metasedimentary, or metavolcanic

Metamorphic rock: formed from igneous or sedimentary rocks by metamorphism

Metamorphism: changes in mineralogy and texture imposed on a rock by elevated pressure and temperature without melting

Migmatite: intimately "mixed" rock with metamorphic and apparent igneous counterparts

Mineral: natural inorganic solid with limited chemical variability and distinctive internal crystalline structure

Moraine: ridge of till left at the glacier front at a "stand"—where the ice remains in fixed position for a long time in response to steady climatic conditions

Moraine, ground: blanket of till with no marked relief formed under the glacier

Mudflat: low-lying strip of muddy ground by the shore, typically more or less submerged by high tide; also tidal flat

Obsidian: volcanic glass formed as very dry lava cools without crystallizing

Organic sediments: made up of organic material, as in coal

Orogeny: long-term, mountain-building event typically accompanied by intense deformation, metamorphism, and igneous activity, and caused by convergence of tectonic plates

Outwash, glacial: glacial deposits washed over, transported, and redeposited by meltwater streams

Overthrust: low-angle thrust fault of large scale, a sheet-like thrust fault

Oxbow lake: crescent-shaped lake formed in an abandoned meander loop which has become separated by a change in the river course

Paragneiss: gneiss formed by metamorphism of sedimentary rock

Pegmatite: very coarse-grained igneous rock, generally of granitic composition, that usually, but not invariably, forms tabular or lenticular bodies

Period: geologic time unit longer than an epoch and a subdivison of an era

Phenocryst: the larger, "floating" crystals in porphyritic igneous rocks

Pillow structure: pillow-like, bulbous masses of lava rock formed when liquid lava flows into water

Piracy, stream: capture of one stream by headward erosion of another with a higher streambed, leaving the lower reach of the captured stream dry

Plastic deformation: taffy-like deformation of rocks, usually under conditions of high temperature and pressure and evident in flowage features

Plate tectonics: a new field of geology which recognizes that the entire lithosphere of the earth, including continents and ocean floors, is made up of a small number of rigid plates of enormous size that are constantly moving about on the underlying asthenosphere

Plucking, glacial: breaking away and removal by glacial ice of large blocks from bedrock

Polish, glacial: smooth surface on bedrock or boulder caused by the polishing action of fine-grained rock waste held by the moving ice

Porphyritic texture: igneous texture with large crystals, called phenocrysts, "floating" in a finer matrix, characteristic of many volcanic rocks

Pothole: rounded hole ground in streambed rock by sand and gravel in swirling eddies

Potsdam Sea: informal name for the shallow sea that covered much of New York in late Cambrian to middle Ordovician time, and which received the original sediment of the Potsdam sandstone

Proglacial lake: lake formed in front of a glacier

Queenston delta: huge wedge of upper Ordovician rocks formed from sediments derived fom the rising ancestral Taconic Mountains in New England and eastern New York

Radiometric clock: any of a number of radioactive elements used to date the rocks that contain them

Rebound, glacial: elastic uplift of the land after the weight of an ice sheet is removed

Recession, glacial: the melting-back of a glacier front primarily caused by climatic warming

Recrystallization: a process in which existing minerals are recrystallized or replaced by new ones without melting during metamorphism

Reef: ridge or mound of coral and other shallow-water marine organisms

Refolding: folding of rocks that have already been folded at least once

Rejuvenation, stream: stream erosion stimulated or renewed, as by uplift of the land or decline of sea level

Resurgence: point at which an underground stream reaches the surface and becomes a surface stream

River terraces: rather flat-surfaced, step-like remnants of former floodplain levels bordering a river that has cut downward

Rock city: city-like arrangement of large, rectangular blocks of resistant caprock that have been separated along joint and bedding planes

Rock record: the preserved rocks

Roof pendant: a downward projection of country rock in the roof of a magma chamber

Salamanca re-entrant: area of southwestern New York that never was covered by glaciers during the Ice Age

Sandstone: sedimentary rock composed of consolidated sand

Sandstone dikes: tabular sandstone bodies that fill fractures in other rocks, formed by invasion of water-saturated sand that later solidified

Sapping: recession of a cliff by wearing away of soft layers that support a harder caprock which then overhangs and breaks off

Scarp: long, more or less continuous cliff or slope separating relatively flat land into two levels

Schist: generally cleaveable metamorphic rock with layers defined by parallel arrangement of platy or prismatic minerals

Scour, glacial: grinding, scraping, gouging, bulldozing action of a glacier

Sedimentary rock: rock formed by consolidation of sediments

Sensitive clay: a clay deposit much like toy "Silly Putty"—firm when undisturbed but easily liquified if shaken or sheared (as by an earthquake)

Shale: sedimentary rock mainly composed of clay or clay-sized material (very fine) and often very thin-bedded

Shear zone: intensely disrupted rock zone marked by numerous closely spaced, parallel shears

Sheeting: closely-spaced rock jointing, generally parallel to and at or near the surface, often caused by pressure release resulting from erosion

Silica: SiO_2 — silicon dioxide

Silicate: compound whose crystal structure contains SiO_4 groups; including most rock-forming minerals

Sill: tabular igneous intrusion between the layers of the intruded rock

Siltstone: sedimentary rock composed of silt, a fine-grained sediment

Sinkhole: pit in the ground formed by solution and collapse of an underground cave

Slate: low-grade metamorphic rock formed from shale and characterized by flat cleavage plates

Soil creep: slow downslope movement of the soil aided by alternate freezing and thawing

Stalactite: icicle-shaped calcium carbonate deposit suspended from a cave ceiling at the site of dripping water

Stalagmite: post-like mass of calcium carbonate rising from the floor of a cave where water drips from the ceiling; usually below a stalactite

Strata: the layers of sediments and sedimentary rocks (also volcanic)

Stratigraphic: term used for layered sediment or sedimentay rock (also for layered, or interlayered, volcanic rocks)

Striae, glacial: linear scratches caused by movement of ice-bound rocks against each other and bedrock

Stromatolites: calcareous algal structures with in-bound sediment and fossil debris, often forming "cabbage heads" and indicative of shallow, near-shore deposition

Summit accordance: alignment of summits over a large region at approximately the same elevation, presumably representing remnants of a former erosional surface that has been dissected

Superposition, Law of: in undisturbed sequences of sedimentary rocks, the oldest layers are on the bottom

Supracrustal: term used for sedimentay and volcanic rocks that overlie the so-called "basement," the older part of the crust upon which they were laid down

Syenite: silica-poor equivalent of granite

Syncline: downfolded rock layers

Talus: fragmentary rock deposit that accumulates at the base of a cliff, mainly by rockfall

Tectonic: term applied to the forces involved in large-scale deformation of the earth's crust

Through-valley: continuous glacially-scoured valley containing a stream divide of glacial deposits

Thrust fault: low angle "push" fault produced by horizontal compression of the crust during convergence of tectonic plates

Tidal flat: extensive, barren, muddy tract that is alternately covered and uncovered by the rise and fall of the tide; also mud flat

Till, glacial: unsorted and unstratified glacial drift deposited directly from the ice without subsequent reworking by streams, usually containing material of a wide range of sizes

Tonawanda, Lake: glacial lake that occupied the shallow basin of the Tonawanda plain between the Niagara and Onondaga scarps after the Wisconsin ice sheet receded from the Niagara scarp

Trough, glacial: broad, elongate, U-shaped valley gouged by a glacier lobe that advanced through a pre-existing stream valley

Type exposure (or locality): the exposure of a particular formation for which it is named, and supposedly typical of the formation

Umlaufbeg: bedrock knob, isolated between pre- and postglacial stream courses

Unconformity: a surface of erosion or non-deposition separating older from younger rocks and constituting a gap in the rock record

Underfit stream: stream much too small to have carved the large features of the valley it occupies, and signifying a drastic diminution of discharge

Valley glacier: linear glacier confined to a mountain valley, also called mountain, or alpine, glacier

Varves: distinctly laminated clays deposited in lakes supplied by glacial meltwater streams, in which the laminations are formed yearly like tree rings

Vermont, Lake: proglacial lake that occupied the Champlain basin in the waning stages of Wisconsin glaciation, and extended well beyond the present shorelines

Wave-cut bench: level or nearly level bedrock surface cut by wave erosion

Wave-cut cliff: cliff produced and maintained by wave undercutting rock or free-standing sediments along the shore

Wisconsin glaciation: the last major glaciation of the Ice Age

SUGGESTED READING

Alt, David, (1982), *Physical Geology*, A Process Approach, Wadsworth Publishing Co., 383 p.

Borst, R. L., (1960), *Rocks and Minerals of New York State*, New York State Museum and Science Service, Educational Leaflet No. 10, 16 p.

Broughton, J. G. et al, (1966), *Geology of New York State*, A Short Account, New York State Museum and Science Service, Educational Leaflet No. 20, 42 p. plus foldouts.

Eicher, D. L. and McAlester, A. L., (1980), *History of the Earth*, Prentice-Hall, Inc., 413 p.

Fisher, D. W., (1962), *Correlation Chart of the Cambrian Rocks in New York State*, New York State Museum and Science Service, Map and Chart Series No. 2.

Fisher, D. W., (1962), *Correlation Chart of the Ordovician Rocks in New York State*, New York State Museum and Science Service, Map and Chart Series No. 3.

Fisher, D. W., (1960), *Correlation Chart of the Silurian Rocks in New York State*, New York State Museum and Science Service, Map and Chart Series No. 1.

Fisher, D. W., Isachsen, Y. W., and Rickard, L. V., (1970), *Geologic Map of New York State* (6 sheets), New York State Museum and Science Service, Map and Chart Series No. 15. Includes a brief tectonic history of New York State.

Fisher, D. W., (1968), *The Genesee River Country*, Information Leaflet, New York State Conservation Dept., L-170, 8 p.

Fisher, D. W., (1968), *Geology of the Plattsburgh and Rouses Point, New York-Vermont Quadrangles*, New York State Museum and Science Service, Map and Chart Series No. 10, 51 p.

Fisher, D. W., Isachsen, Y. W., and Whitney, P. R., (1980), *New Mountains from Old Rocks*, the Adirondacks, the Geology of the Lake Placid Region, New York State Geological Survey, Educational Leaflet.

Flint, R. F. et al, (1959), *Glacial Geology of the United States East of the Rocky Mountains* (map), Geological Survey of America.

Isachsen, Y. W., (1980), *Continental Collisions and Ancient Volcanoes*, The Geology of Southeastern New York, New York State Geological Survey, Educational Leaflet 24, 15 p.

Isachsen, Y. W. and McKendree, W. G., (1977), *Preliminary Brittle Structures Map of New York* (4 sheets), New York State Museum and Science Service, Map and Chart Series No. 31A.

New York State Geological Association Field Trip Guidebooks
(1925-1984, annual except for 1942-45). These are available in
many libraries in New York State.

Rickard, L. V., (1964), *Correlation Chart of the Devonian Rocks in
New York State*, New York State Museum and Science Service,
Map and Chart Series No. 4.

Roseberry, C. R., (1982), *From Niagara to Montauk*, the Scenic
Pleasures of New York State, The State University of New York
Press, 344 p.

Schuberth, C. J., (1968), *The Geology of New York City and Environs*,
The American Museum of Natural History, The Natural History
Press, 304 p. with map in pocket.

Staehle, H. C. et al, (no date given, but 1970s), *Getting Acquainted
with the Geological Story of the Rochester and Genesee Valley
Areas*, Rochester Academy of Science, 84 p.

Tesmer, I. H., ed., authors: Bastedo, J. C., Brett, C. E., Fisher, D. W.,
Kilgour, W. J., Krajewski, J. L., Liberty, B. A., Terasmae, J.,
(1981), *Colossal Cataract*, the Geological History of Niagara
Falls, The State University of New York Press, 219 p. with map in
pocket.

Van Diver, B. B., (1976), *Rocks and Routes of the North Country*, New
York, W. F. Humphrey Press, 205 p.

Van Diver, B. B., (1980), *Upstate New York*, Geology Field Guide,
Kendall/Hunt Publishing Co., 276 p.

Von Engeln, O. D., (1961), *The Finger Lakes Region*, Its Origin and
Nature, Cornell University Press, 156 p.

Index

398

400